RESIDUE REVIEWS

VOLUME 34

RESIDUE REVIEWS

Residues of Pesticides and Other
Foreign Chemicals in Foods and Feeds

RÜCKSTANDS-BERICHTE

Rückstände von Pestiziden und anderen
Fremdstoffen in Nahrungs- und Futtermitteln

Editor
FRANCIS A. GUNTHER

Assistant Editor
JANE DAVIES GUNTHER

Riverside, California

VOLUME 34

SPRINGER-VERLAG
BERLIN • HEIDELBERG • NEW YORK
1971

ISBN-13: 978-1-4615-8472-8 e-ISBN-13:978-1-4615-8470-4

DOI: 10.1007/ 978-1-4615-8470-4

Library of Congress Catalog Card Number 62-18595.

Preface

That residues of pesticide and other "foreign" chemicals in foodstuffs are of concern to everyone everywhere is amply attested by the reception accorded previous volumes of "Residue Reviews" and by the gratifying enthusiasm, sincerity, and efforts shown by all the individuals from whom manuscripts have been solicited. Despite much propaganda to the contrary, there can never be any serious question that pest-control chemicals and food-additive chemicals are essential to adequate food production, manufacture, marketing, and storage, yet without continuing surveillance and intelligent control some of those that persist in our foodstuffs could at times conceivably endanger the public health. Ensuring safety-in-use of these many chemicals is a dynamic challenge, for established ones are continually being displaced by newly developed ones more acceptable to food technologists, pharmacologists, toxicologists, and changing pest-control requirements in progressive food-producing economies.

These matters are of genuine concern to increasing numbers of governmental agencies and legislative bodies around the world, for some of these chemicals have resulted in a few mishaps from improper use. Adequate safety-in-use evaluations of any of these chemicals persisting into our foodstuffs are not simple matters, and they incorporate the considered judgments of many individuals highly trained in a variety of complex biological, chemical, food technological, medical, pharmacological, and toxicological disciplines.

It is hoped that "Residue Reviews" will continue to serve as an integrating factor both in focusing attention upon those many residue matters requiring further attention and in collating for variously trained readers present knowledge in specific important areas of residue and related endeavors; no other single publication attempts to serve these broad purposes. The contents of this and previous volumes of "Residue Reviews" illustrate these objectives. Since manuscripts are published in the order in which they are received in final form, it may seem that some important aspects of residue analytical chemistry, biochemistry, human and animal medicine, legislation, pharmacology, physiology, regulation, and toxicology are being neglected; to the contrary, these apparent omissions are recognized, and some pertinent manuscripts are in preparation. However, the field is so large and the interests in it are so varied that the editors and the Advisory Board earnestly solicit suggestions of topics and authors to help make this international book-series even more useful and informative.

"Residue Reviews" attempts to provide concise, critical reviews of timely advances, philosophy, and significant areas of accomplished or needed endeavor in the total field of residues of these chemicals in foods, in feeds, and in transformed food products. These reviews are either general or specific, but properly they may lie in the domains of analytical chemistry and its methodology, biochemistry, human and animal medicine, legislation, pharmacology, physiology, regulation, and toxicology; certain affairs in the realm of food technology concerned specifically with pesticide and other food-additive problems are also appropriate subject matter. The justification for the preparation of any review for this book-series is that it deals with some aspect of the many real problems arising from the presence of residues of "foreign" chemicals in foodstuffs. Thus, manuscripts may encompass those matters, in any country, which are involved in allowing pesticide and other plant-protecting chemicals to be used safely in producing, storing, and shipping crops. Added plant or animal pest-control chemicals or their metabolites that may persist into meat and other edible animal products (milk and milk products, eggs, etc.) are also residues and are within this scope. The so-called food additives (substances deliberately added to foods for flavor, odor, appearance, etc., as well as those inadvertently added during manufacture, packaging, distribution, storage, etc.) are also considered suitable review material.

Manuscripts are normally contributed by invitation, and may be in English, French, or German. Preliminary communication with the editors is necessary before volunteered reviews are submitted in manuscript form.

Department of Entomology
University of California
Riverside, California
December 10, 1970

F.A.G.

Table of Contents

Determination of organophosphorus pesticide residues in fruits and vegetables on the Swedish market from 1964 to 1968

By

SIV RENVALL* and MALIN ÅKERBLOM*

Contents

I. Introduction

International food trade has been accelerated by rapid enlargement of trading areas, improved transportation, and food technology. Large quantities of fresh fruit and fresh vegetables are imported into Sweden from all parts of the world. The chemical treatment of produce from different countries is difficult to survey.

---

* National Institute of Plant Protection, 171 07 Solna, Sweden.

Considerable attention has recently been focused on the possible health hazards of pesticide residues in foods, and the Food and Agriculture Organization (FAO) and World Health Organization (WHO) have published a series of reports dealing with general principles of pesticide residue safety in food. For some pesticides tolerances and temporary tolerances have been recommended (FAO/WHO 1968 a and b, 1969 a and b).

Many investigations have been made on pesticide residues in foods ready for consumption, including data obtained by total diet studies (e.g., Abbott *et al.* 1969, Coffin 1964, Duggan and Weatherwax 1967, Harries *et al.* 1969, Krause and Kirchhoff 1969, Thompson and Hill 1969, Westöö *et al.* 1969). Most of these investigations deal with organochlorine pesticides. For some years the organophosphorus pesticides have been used on an increasing scale as insecticides and acaricides in agriculture and in orchards. They tend to replace the organochlorine pesticides, which have been in use for many years. In total diet studies, the presence of organophosphorus pesticides has been reported less frequently than the presence of organochlorine compounds. This is probably partly due to the shorter persistence of these compounds, and to the lower dosage necessary for pest control, but may also be due to analytical problems associated with these compounds, which undergo numerous chemical alterations after application to the crop. Some of these intermediate compounds are more toxic than the parent compounds.

The purpose of the present investigation was to examine residues of organophosphorus pesticides in domestic and imported fruit and vegetables of unknown agricultural history marketed in Sweden. About 2,400 samples were collected and analysed from 1964 to 1968.

II. Methods

Methods for the determination of residues of organophosphorus pesticides in foods were reviewed in 1967 by Abbott and Egan. The methods presented in this investigation were made as simple as possible to give a survey of the presence of the most common organophosphorus pesticides in fruit and vegetables.

A cholinesterase inhibition method has been used as a screening test to sort out samples without detectable amounts of pesticides (Archer 1963, Gage 1961). The Warburg manometric technique was used (Augustinsson 1957). In most samples with cholinesterase inhibition the pesticides were identified and quantitatively determined by gas-liquid chromatography. In samples with more than 0.1 mg./kg., the results were confirmed by thin-layer chromatography. Quantitative determination of metabolites is not included in the investigation; the metabolites were observed only as an unidentified cholinesterase inhibition effect.

The individual steps of the analytical procedure were:

Extraction of the chopped sample with dichloromethane.

Screening of the dichloromethane ("crude") extract for cholinesterase inhibition.

Cleanup of the crude extract on an alumina column. Elution with petroleum ether, fraction I, and thereafter with petroleum ether/acetone, fraction II.

Determination of the cholinesterase inhibition of fraction II, and in some cases, of fraction I.

Gas chromatographic analysis on two different columns with the phosphorus detector.

Thin-layer chromatography for verification of identity of pesticides observed.

a) Extraction procedure

From a fresh sample of 0.5 to 1.5 kg. of fresh crop a finely chopped aliquot of 100 g. (50 g. of lettuce or spinach) was taken. It was macerated in a Turmix blendor with 200 ml. of dichloromethane (purum, redistilled, b.p. 39° to 41° C.) for two-to-three minutes. The slurry was transferred to a stoppered flask and was left to the next day at room temperature. The slurry was filtered and the crude dichloromethane extract was dried by passage through a 3-cm. column (i.d. 1.5 cm.) of anhydrous sodium sulphate (A.R.). A reference aliquot of 50 or 100 g. of every chopped sample was deep-frozen.

Many solvents have been reported in the literature for the extraction of organophosphorus pesticide residues from crops. LAWS and WEBLEY (1961) recommended dichloromethane for its favourable extraction characteristics for both organochlorine and organophosphorus pesticide residues including metabolites. This solvent has been used in the present investigation. The effectiveness of one extraction with dichloromethane has been tested against repeated extractions of samples containing residues with satisfactory results. It was observed that on storage of dichloromethane extracts of samples with residues in brown bottles at room temperature (25° C.), most of the detected pesticides were very stable. The concentration of ethion and parathion in these extracts was the same after one year. The same stability was not found in extracts purified by alumina column chromatography.

b) Cleanup procedure

A chromatographic tube (1.5 cm. i.d. and 40 cm. long) was packed with ten g. (5.5 cm.) of aluminum oxide (activity grade II–III, BROCKMANN) and a top layer of two cm. of anhydrous sodium sulphate. The column was washed and pre-wetted with 50 ml. of petroleum ether (purum, redistilled, b.p. 40° to 50° C.). An aliquot of 100 ml. of crude extract was evaporated just to dryness in a vacuum evaporator at 30° to 40° C. The residue was transferred to the column with three ml. of petroleum ether. The evaporation flask and the sides of the tube were washed with 3 + 3 ml. of petroleum ether. The column was eluted with 50 ml. of petroleum ether (fraction I), then with 100 + 25 ml. of three percent v/v acetone (purum, redistilled, b.p. 55.5° to 56.5° C.) in petroleum ether (fraction II:1 and II:2).

With gas chromatographic determination of organophosphorus pesticides with a specific phosphorus detector, the cleanup of crop extracts does not need to be very thorough. The absorption of pigments and waxes on a chromatographic column is, however, advisable. For thin-layer chromatography a cleanup is necessary.

c) *Determination of cholinesterase inhibition*

The Warburg manometric technique was used for assaying cholinesterase activity.

1. Reagents.

Standard solutions of pesticides: A concentration of one μg./ml. (in ethyl acetate or petroleum ether) is convenient. The purities of the pesticides involved are listed in Table I.

Chlorinesterase solutions: (a) Purified preparation from human blood serum, freeze-dried,[1] 40 mg./100 ml. of buffer solution. (b) Bovine cholinesterase (pseudo) B grade, 6.3 units/mg. (Calbiochem, Lucerne, Switzerland), 7.5–10.0 mg./100 ml. of buffer solution. The enzyme solutions are deep-frozen in portions of 7–8 ml., which is the amount needed for a Warburg set. Enzyme (a) was used from 1964 to 1965 and (b) from 1966 to 1968.

Bromine water: 0.2 ml. of glass-distilled water saturated with bromine was diluted to 50 ml. Fresh solution was prepared every two weeks.

Buffer solution: According to AUGUSTINSSON (1957), *i.e.,* 0.15 *M* NaCl, 0.025 *M* $NaHCO_3$, and 0.04 *M* $MgCl_2 \cdot 6H_2O$ in glass-distilled water, pH 7.8. Fresh solution was prepared every day.

Substrate: 0.0666 g. of acetylcholine chlorhydrate (Merck, Darmstadt, Germany) dissolved in five ml. of buffer solution just before use.

2. Apparatus.

Warburg apparatus (B. Braun, Melsingen, Germany) with 14 manometers. The flasks have one side bulb and a total capacity of approximately 15 ml. each.

3. Procedure.

One ml. of the crude or purified crop extract was added to the main compartment of the Warburg flask. The solvent was evaporated by a stream of air and the residue was treated with dilute bromine water, 0.2 ml. for crude and 0.1 ml. for purified extracts. Exactly 0.5 ml. of enzyme

---

[1] Kindly donated by AB Kabi, Stockholm, Sweden.

Table I. *Cholinesterase inhibition*[a] *and toxicity of various organophosphorus pesticides*

Pesticide	Source[b]	Purity (%)	Molar conc. for 50% enzyme inhibition	LD_{50} rats (mg./kg.)
azinphos-methyl	2	99	2.1×10^{-8}	7–13
bromophos	4	90	4.6×10^{-9}	3,750–5,180
carbophenothion	10	94.6	4.9×10^{-9}	7–30
carbophenothion sulfone	10	80–90	4.4×10^{-9}	—
carbophenothion sulfoxide	10	60–70	1.9×10^{-9}	—
carbophenothion *O*-analogue	10	89	5.1×10^{-9}	—
carbophenothion-methyl	10	83.4	1.1×10^{-8}	—
chlorfenvinphos	9	85.6 α-isomer 10.1 β-isomer	3.7×10^{-9}	10–135
demeton-*S*-methyl	2	92.4	4.3×10^{-6}	40
diazinon	7	93	5.5×10^{-9}	300–600
dichlofenthion	3	90–100[c]	2.1×10^{-8}	250
dichlorvos	3	90–100[c]	7.5×10^{-8}	25–30
dimethoate	4	95	2.9×10^{-6}	200–300
EPN	3	90–100[c]	1.0×10^{-8}	8–17
ethion	3	90–100[c]	8.7×10^{-9}	13–34
fenchlorphos	5	44[d]	1.0×10^{-8}	1,000–3,000
fenitrothion	2	99.6	1.2×10^{-7}	250–670
malathion	1	99.7	5.0×10^{-7}	1,400–1,900
mecarbam	6	90	2.0×10^{-8}	15
mevinphos	3	90–100[c]	1.5×10^{-7}	3–5
naled	6	90	4.4×10^{-9}	430
parathion	3	90–100[c]	1.1×10^{-8}	3–6
parathion-methyl	3	90–100[c]	2.5×10^{-7}	12–16
phorate	1	91.5	9.0×10^{-8}	2–3
phosalone	7	30[d]	9.1×10^{-10}	120–170
phosphamidon	6	90	1.1×10^{-7}	15
thiometon	8	25[d]	1.4×10^{-6}	100
trichloronat	2	95	2.0×10^{-9}	37

[a] Cholinesterase inhibition was tested by the manometric method with bovine cholinesterase at 25°C; thioorganophosphorus compounds were activated by bromine water except for dimethoate which was activated by irradiation with UV light.

[b] Sources: 1 = American Cyanamid Co., Gosport, Hants, England; 2 = Farbenfabriken Bayer AG., Leverkusen, Germany; 3 = Chemical Service Inc., Media, Pa., U.S.A.; 4 = Gullvik AB, Malmö, Sweden; 5 = Klärre & Co., Stockholm, Sweden; 6 = Nordisk Alkali AB, Malmö, Sweden; 7 = Philips AB, Stockholm, Sweden; 8 = Sandoz AG, Basel, Switzerland; 9 = Shell Chemical Co., Sittingbourne, England; and 10 = Stauffer Chemical Co., Richmond, Calif., U.S.A.

[c] Ten percent solution in analytical grade benzene.

[d] Formulated substance.

solution and 2.0 ml. (2.1 ml.) of buffer solution were added to the main compartment and the side bulb was filled with 0.3 ml. of substrate solution. The total volume of the reaction mixture was thus 3.00 ml. In every series there were one thermobarometer with 3.00 ml. of buffer solution for corrections due to alterations in temperature and atmospheric pressure and two flasks with enzyme and substrate in buffer solution to get the total enzyme activity. To one of these controls bromine water was added. The flasks were connected to the manometers and the gas mixture (95 percent N_2 and five percent CO_2) was bubbled through. The apparatus was placed in the water thermostat to attain 25°C. equilibrium before the contents of the flasks were mixed. The incubation time for the test sample and the enzyme was about one hour and then the substrate was added. The manometers were read immediately before mixing enzyme and substrate at zero time and subsequently every ten minutes for one hour.

Suitable amounts of standard solutions of pesticides were tested for inhibition effect in the same manner as purified extracts.

The amount of CO_2 expressed in μl. evolved from the bicarbonate buffer by the acetic acid produced during the reaction was plotted against time in minutes. The slope of the straight line thus obtained was used as a measure of the enzyme activity.

The 30-minute value from the curve was used in calculating the enzyme inhibition caused by the test sample. The results were corrected for nonenzymic ester hydrolysis of the substrate. The inhibition effect was expressed in percent of the enzyme activity of the control.

Many organophosphorus pesticides are very strong inhibitors of cholinesterase. Thionophosphates with weak inhibition effect can be converted into more potent cholinesterase inhibitors by an oxidation with diluted bromine water (FALLSCHEER and COOK 1956) or by irradiation with ultraviolet light (MITCHELL *et al.* 1968).

Twenty-six organophosphorus pesticides were tested with the manometric method for determination of the cholinesterase inhibition effect. Table I gives the amount of the pesticides which gives 50 percent inhibition (I_{50} values) of bovine cholinesterase at 25° C. All pesticides except dimethoate and non-thiophosphates (chlorfenvinphos, naled, mevinphos, and phosphamidon) were oxidized with bromine water to oxygen-analogues. Dimethoate was oxidized by ultraviolet irradiation. Table I also shows the approximate acute oral LD_{50} values in rats (JONES *et al.* 1968) and the purity of the chemicals used; most of the chemicals were technical grade. The names of the pesticides given in the tables are ISO-recommended common names and proposed common names (ISO 1959–1969).

To get reproducible values of the concentration of a pesticide in a crop by the cholinesterase inhibition method the extract must be purified from co-extractives which can influence the enzyme. Table II summarizes I_{50} values of pesticides in the presence of purified extract (fraction II:1 from the cleanup step) of different crops. Thionophosphates were treated with bromine water (dimethoate was irradiated with ultraviolet light). The table shows that the I_{50} values are

Table II. *Cholinesterase inhibition effect of organophosphates and activated thiophosphates in the presence of extracts of different crops*[a]

Pesticide	Pesticide (µg. in 3 ml.) which gives 50% enzyme inhibition in presence of extracts of						
	No extract	Apples (0.5 g.)	Lettuce (0.25 g.)	Pears (0.5 g.)	Potatoes (0.5 g.)	Tomatoes (0.5 g.)	Cucumbers (0.5 g.)
azinphos-methyl	0.02	0.04	0.02	0.04	0.02	—	—
bromophos	0.005	0.1	0.01	—	0.01	—	—
carbophenothion	0.005	0.03	0.003	0.005	0.003	0.005	—
chlorfenvinphos	0.004	0.002	0.002	—	0.002	—	—
demeton-*S*-methyl	3	3	3	—	3	—	—
diazinon	0.005	0.005	0.0005	—	0.0005	—	0.002
dichlofenthion	0.02	0.10	0.02	—	0.02	—	—
dichlorvos	0.05	0.05	0.03	—	0.03	—	—
dimethoate	2	2	2	—	2	—	—
EPN	0.01	0.1	0.03	—	0.03	—	—
ethion	0.01	0.01	0.002	0.01	0.002	—	—
fenchlorphos	0.01	0.2	0.01	—	0.02	0.1	—
fenitrothion	0.1	0.2	0.03	0.1	0.06	—	—
malathion	0.5	2	1	—	0.8	—	—
mecarbam	0.02	0.03	0.02	—	0.02	—	—
mevinphos[b]	0.1	0.1	0.1	—	0.1	—	—
naled	0.005	0.02	0.02	—	0.02	—	—
parathion[c]	0.01	0.02	0.005	0.01	0.005	0.02	0.01
parathion-methyl	0.2	0.6	0.3	0.5	0.3	—	—
phorate	0.07	0.2	0.1	0.2	0.1	—	—
phosalone	0.001	0.003	0.001	—	0.001	—	—
phosphamidon	0.1	0.2	0.2	—	0.2	—	—
thiometon	1	2	1	—	1	—	—
trichloronat	0.002	0.03	0.003	—	0.005	—	—

[a] The pesticides were added to extracts which had been purified on an alumina column.
[b] Broccoli (0.5 g.), 0.1 µg.
[c] Cherries (0.5 g.), 0.01 µg.; spinach (0.25 g.), 0.01 µg.

higher in the presence of apple extract than in presence of other tested crop extracts, particularly lettuce. If there is only one pesticide in an extract the figures can be used for approximate determination of its concentration in the crop.

d) Gas chromatography (GC)

Wilkens Aerograph gas chromatograph model no. 204 with the Aerograph phosphorus detectors was used (Hartmann 1966). Borosilicate glass columns, 5′ × ⅛″ o.d., were silanized by treatment with five percent dichlorodimethylsilane in toluene, packed with four percent w/w SE-30 on Chromosorb G AW-DMCS 70/80 mesh or with two percent w/w QF-1 on Chromosorb G AW-DMCS 70/80 mesh, and conditioned 60 to 72 hours before use. The unit was operated isothermally with the oven temperature between 175° and 215° C. (see later). The temperature of the glass-lined injection ports and of the detectors were 5° to 10° C. above the oven temperature. Nitrogen filtered through a molecular sieve (5Å, Aerograph) was used as carrier gas at about 25 ml./minute. The cesium bromide pellet on top of the quartz tip of the phosphorus detector was sanded once a week to present a clean surface to the flame. Cylinder air and hydrogen feeding the flame were filtered through molecular sieves (5Å, Aerograph). The air flow rate was 170 ml./minute. To get optimum conditions the hydrogen flow rate was adjusted to give a background current of 2.7×10^{-9}A (50 percent deflection of full scale on a one-mV recorder with the electrometer set at range one, attenuation 16); this corresponded to a hydrogen flow rate of 10–14 ml./minute.

Ten ml. of purified extract was concentrated by a stream of air to 0.5 ml. Six μl. of the concentrated solution, corresponding to 60 mg. of sample, was injected; these injections sometimes blew out the flame, but this did not disturb the analysis if the flame was immediately relit. As the phosphorus detector is highly specific for phosphorus compounds it permits crude extracts to be injected directly into the gas chromatograph. This has been done with crude extracts where the cholinesterase inhibition effect was not recovered in the purified extracts. In these cases, five ml. of the crude extract was taken just to dryness and redissolved in 0.5 ml. of ethyl acetate and six μl. was injected. However, a cleanup is preferable so that the columns are not spoilt. For identification every sample giving peaks was run on both columns (the more polar QF-1 and the less polar SE-30).

In Table III retention data on the two columns for 31 pesticides are presented. Table III also gives approximate values for the peak heights of the pesticides relative to parathion for both columns between 175° and 215° C. with the phosphorus detector. The minimum detectable amount of parathion was 0.1 ng. (*i.e.*, 0.002 mg./kg. of test sample).

Difficulties have been encountered when azinphos-methyl was analysed; this substance, therefore, might be present in more samples than denoted. Dichlorvos has also been difficult to detect since it has a short retention time and the solvent peak very often interferes.

Some of the organophosphorus pesticides (e.g., ethion, parathion) are also detectable with the electron capture detector. Comparative investigations have been made with electron capture and phosphorus detectors for determination of ethion and parathion. With the electron capture detector, however, the presence of co-extractives, among them organochlorine pesticides, may disturb the determination. Figure 1 shows chromatograms of a pear extract analysed with both phosphorus and electron capture detectors. As can be seen, the electron capture detector needs a more thorough cleanup.

Table III. *Gas chromatography (GC) and thin-layer chromatography (TLC) data for organophosphorus pesticides with all data related to parathion*[a]

Pesticide	GC					TLC
	Relative retention time				Relative peak height	Relative R_f value
	SE-30		QF-1			
	175° C.	211° C.	175° C.	211° C.		
azinphos-methyl	—	4.9	—	4.2	—[b]	0.20
bromophos	1.13	—	0.53	—	0.6	1.39
carbophenothion	3.20	2.60	1.33	—	0.3	1.47
carbophenothion-methyl	—	2.10	1.06	—	0.3	1.22
carbophenothion-*O*-analogue	—	2.10	1.77	—	0.3	0.18
carbophenothion sulfone	—	4.5	—	4.6	0.05	0.74
carbophenothion sulfoxide	—	4.9	—	4.4	0.05	0.28
chlorfenvinphos	1.34	—	0.98	—	0.5	0.17
demeton-*S*-methyl	0.26	—	0.24	—	1	0.08
diazinon	0.51	—	0.20	—	2.3	0.78
dichlofenthion	0.67	—	0.32	—	1.2	1.47
dichlorvos	0.07	—	0.06	—	6.8	0.17
dimethoate	0.40	—	0.52	—	1	0.03
EPN	—	3.8	—	3.2	0.2	1.00
ethion	2.78	2.20	1.40	—	0.5	1.28
fenchlorphos	0.80	—	0.39	—	1	1.40
fenitrothion	0.86	—	0.87	—	1	0.80
malathion	0.91	—	0.74	—	1	0.57
mecarbam	1.30	1.25	1.09	—	0.7	0.66
mevinphos	0.13	—	0.17	—	1.4	0.02
naled	0.31	—	0.23	—	0.6	0.02
parathion	1.00	1.00	1.00	1.00	1	1.00
parathion-methyl	0.69	0.80	0.77	—	0.7	0.69
phorate	0.35	—	0.13	—	2.5	1.49
phorate sulfone	0.96	—	1.32	—	0.5	—
phorate sulfoxide	0.98	—	1.18	—	0.3	—
phorate-*O*-analogue	0.24	—	0.21	—	2	—
phorate-*O*-analogue sulfone	0.73	—	1.53	—	0.06	—
phorate-*O*-analogue sulfoxide	0.77	—	1.44	—	0.03	—
phosalone	—	4.9	—	4.2	0.06	0.73
trichloronat	1.17	—	0.47	—	1	1.68

[a] GC: Retention time of parathion on SE-30 at 175° C. was 10.5 minutes and at 211° C. 3.3 minutes; on QF-1 at 175° C. it was 7.5 minutes and at 211° C. 2.3 minutes. Peak heights refer to analysis with the phosphorus detector. TLC: R_f values were recorded on silica gel plates with *n*-hexane/ethyl acetate (4 + 1, v/v) as the developing system. For parathion the R_f value was 0.30 to 0.39.

[b] Non-reproducible results.

e) *Thin-layer chromatography*

Desaga standard equipment was used to prepare thin-layer chromatography plates of silica gel G, 0.25 mm., activated at 110° C. for two hours. Ten ml. of purified extract was taken to dryness in a stream of air and was then redissolved in 0.2 ml. of *n*-hexane and spotted on the plates. Of standard pesticides, 0.2–2 μg. were spotted. The plates were developed with *n*-hexane/ethyl acetate, 4 + 1 v/v. The solvent front was allowed to migrate 12 cm. above the spotting line. The plates were dried at room temperature before spraying.

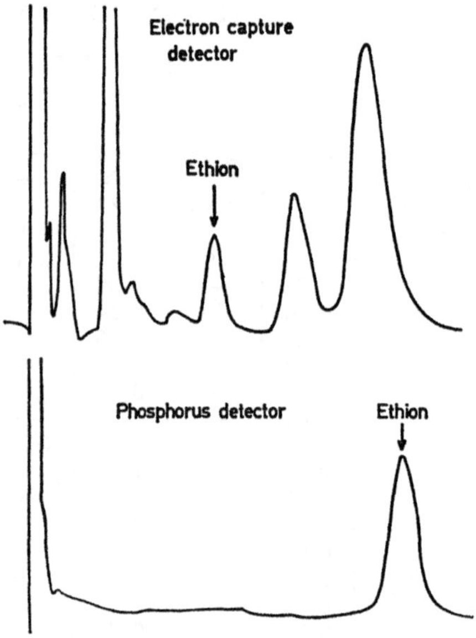

Fig. 1. Response of electron capture detector compared with that of phosphorus detector after chromatography of 0.07 mg./kg. of ethion in a pear extract; for electron capture a QF-1 column and for the phosphorus detector an SE-30 column was used

Chromogenic agents according to KOVACS (1964) were used as spray reagents for detection of thiophosphate pesticides:

(a) 0.1 g. of tetrabromophenolphthalein ethyl ester potassium salt in ten ml. of acetone diluted to 50 ml. with ethanol (99 percent).

(b) 1.0 g. of $AgNO_3$ (A.R.) dissolved in five ml. of distilled water and diluted to 200 ml. with ethanol (96 percent). One drop of H_2O_2 solution, 30 percent, was added and the solution was stored in a dark bottle.

(c) 5.0 g. of citric acid (A.R.) dissolved in 50 ml. of distilled water and diluted to 100 ml. with ethanol (96 percent).

The developed plates were sprayed with dye solution (a), then with the $AgNO_3$ solution (b) and after two minutes with citric acid solution (c). After spraying the thiophosphate pesticides immediately appeared as vivid blue or purple spots against a yellow background.

For the detection of organophosphorus pesticides which do not contain sulphur the chromogenic reagents according to WATTS (1965) were used:

(a) 0.25 percent 4-(*p*-nitrobenzyl)-pyridine (NBP) in redistilled acetone.

(b) 10 percent tetraethylenepentamine in redistilled acetone, prepared fresh daily. (The acetone was refluxed with one g. of $KMnO_4$/liter for one hour and distilled.)

The developed chromatograms were sprayed with solution (a). After the acetone had evaporated, the plates were heated at 150° C. for 15 minutes. The plates were allowed to cool for ten minutes and were then sprayed lightly with solution (b) until development of maximum spot intensity. The spots were blue against a white background.

With the spray solutions according to WATTS phosphate as well as thiophosphate pesticides can be detected. To get good results, the crop extracts, however, have to be very clean, otherwise the coextractives interfere too much. The KOVACS' solutions are preferable, because they give better results for crop extracts and residues of thiophosphate pesticides in crops down to 0.1 mg./kg. can be detected.

Table III gives the R_f values relative to parathion for the pesticides. The R_f value for parathion is 0.30–0.39.

f) Recovery experiments

Aliquots of standard solutions of pesticides were added either directly to the alumina column or to chopped crops before maceration. The samples were cleaned up and analysed by gas chromatography according to the methods described above. Recoveries are given in Table IV.

In recovery experiments starting on column an addition of five to 50 μg. of pesticide in petroleum ether was made, which corresponds to 0.1 to 1.0 mg./kg. of crop in the method used. The mean values of the recoveries were 89 to 99 percent. Pesticides which are mentioned in Table III but not in Table IV could not be eluted from the alumina column by the present method.

The pesticides added to crops were 0.1 to 0.5 mg./kg. of crop. The addition was usually made directly before maceration, with mean recovery values of 70 to 106 percent. To some apple, pear, and spinach samples the pesticides were added 30 minutes before the maceration in dichloromethane. The apples and pears had been chopped before this addition, and the spinach had been taken from frozen, minced produce. The recoveries were then reduced, particularly for diazinon

Table IV. *Percent recoveries of organophosphorus pesticides from the cleanup plete analysis with addition of 0.1–0.5 mg./kg. to untreated crop sample before*

Pesticide	Recoveries (%)[a] of			
	Column	Apples	Cauliflower	Cucumbers
bromophos	95 ± 6.7 (6)	—	91 (1)	—
carbophenothion	98 ± 3.3 (6)	93 ± 13.2 (5)	—	—
carbophenothion-methyl[c]	91 ± 6.9 (5)	93 ± 13.5 (4)	—	—
diazinon	97 ± 9.6 (10)	87 ± 4.4 (5)	—	97 ± 9.0 (4)
	—	45,46,49 (3)[d]	—	—
dichlofenthion	99 ± 3.1 (5)	—	72,78,85 (3)	—
EPN	—	—	—	86 (1)
ethion	97 ± 10.2 (8)	95 ± 7.9 (5)	69,75 (2)	—
	—	81 ± 5.8 (5)[d]	—	—
fenchlorphos[b]	93 ± 5.5 (5)	88 ± 5.7 (4)	—	—
fenitrothion[c]	89 ± 6.3 (5)	—	—	—
	—	—	—	—
parathion	95 ± 12.9 (8)	93 ± 5.1 (5)	65,65,80 (3)	96 ± 12.0 (4)
	—	86 ± 22.5 (5)[d]	—	—
parathion-methyl[c]	89 ± 12.2 (6)	86 ± 7.9 (5)	—	86 ± 14.5 (4)
phorate	96 ± 7.0 (5)	—	—	—
trichloronat[b]	98 ± 3.6 (5)	—	114 (1)	—

[a] Mean values ± standard deviation. Values in brackets denote the number of analyses run.
[b] The pesticide was eluted in fractions I and II:1.

in apples and for fenitrothion and parathion in spinach. Few such recovery experiments were carried out, but they indicate that pesticides may be rapidly metabolized in certain chopped crops.

III. Results and discussion

a) Collection and origin of samples

The samples analysed have been purchased in shops and at wholesalers at different places in Sweden, and some of the imported crops were also obtained at the frontier inspection. The number of samples amount to 2,396, taken from July 1964 to December 1968. The domestic produce was taken mainly from May to December, and the imported wares mainly from October to June. Samples from each crop were collected throughout the season according to what was commercially available. Table V summarizes species and origin of the produce and the number of samples.

step with 5–50 μg. of pesticide added to the alumina column and from the com- maceration

pesticides added to

Horseradish	Lettuce	Pears	Potatoes	Spinach	Tomatoes
—	—	—	—	—	—
—	—	101 ± 10.8 (5)	—	—	89 ± 11.3 (5)
—	—	101 ± 8.4 (4)	—	—	91 ± 7.2 (4)
96,107,110 (3)	99 ± 5.4 (5)	90 ± 9.1 (5)	—	—	—
—	—	—	—	—	—
—	—	—	—	—	—
—	68 (1)	66 (1)	—	—	—
96 (1)	106 (1)	98 ± 7.9 (5)	70,97 (2)	—	—
—	—	64,75 (2)[d]	—	—	—
—	87 (1)	—	—	—	102 ± 4.7 (4)
—	—	86 ± 16.0 (5)	—	98 (1)	79 (1)
—	—	—	—	26 (1)[d]	—
—	98 ± 2.3 (5)	95 ± 1.9 (5)	—	92 (1)	102 ± 14.4 (5)
—	—	65,66 (2)[d]	—	40 (1)[d]	—
—	83 ± 5.4 (4)	90 ± 3.0 (4)	—	—	—
—	—	—	104 (1)	—	—
103 ± 9.9 (5)	—	—	—	—	—

[c] The pesticide was eluted in fractions II:1 and II:2.
[d] The pesticide was added to the crop 30 minutes before extraction.

Some of the samples of apples and pears have also been analysed for organochlorine pesticides by the National Institute of Public Health, Stockholm, Sweden (Westöö *et al.* 1968).

b) Results of screening with cholinesterase

Figure 2 (Group A) summarizes the results of the screening test on crude extracts of samples of 13 different crops. The results were arbitrarily divided into three groups: with no or weak enzyme inhibition, 0–29 percent, with medium inhibition, 30 to 50 percent, and with strong inhibition, >50 percent. In samples with weak inhibition, the occurrence of organophosphorus pesticides is doubtful because even extracts of untreated samples, especially of lettuce and cucumbers, may show a slight inhibition. About 78 percent (2,150) of the samples had no or weak enzyme inhibition, 11 percent had medium, and 11 percent had strong enzyme inhibition. The strong enzyme inhibition was found more frequently in some crops than in others. Thus, spinach and domestic plums always had no or weak inhibition and beans,

Table V. *Crop species, origin, and number of samples*

Country	Produce																	
	Apples	Beans	Broccoli	Carrots	Cauliflower	Cherries	Cucumbers	Grapes	Lettuce	Peaches	Pears	Peas	Plums	Potatoes	Spinach	Tomatoes	White cabbage	Sum
Argentina	30	—	—	—	—	—	—	17	—	—	33	—	—	—	—	—	—	80
Australia	16	—	—	—	—	—	—	—	—	—	9	—	—	—	—	—	—	25
Belgium	—	—	—	—	—	—	1	1	6	—	1	—	—	—	—	2	—	11
Bulgaria	1	—	—	—	3	1	4	14	—	1	—	—	—	1	—	12	1	38
Canada	9	—	—	—	—	—	—	—	—	—	3	—	—	—	—	—	—	12
Chile	11	—	—	—	—	—	—	1	—	—	5	—	—	—	—	—	—	17
China	3	—	—	—	—	—	—	—	—	—	—	—	—	—	—	—	—	3
Denmark	1	—	—	20	—	—	3	—	1	—	—	—	—	1	—	—	3	29
Finland	—	—	—	—	—	—	—	—	—	—	—	—	—	—	—	1	—	1
Formosa	—	—	—	—	—	—	—	—	—	—	—	—	1	—	—	—	—	1
France	9	—	—	—	7	27	—	1	2	—	3	—	—	—	—	—	2	51
Germany	5	—	—	—	—	1	—	—	—	—	—	—	—	—	—	—	1	7
Hungary	18	1	—	—	—	5	2	—	2	—	—	—	1	—	—	—	4	33
Israel	—	—	—	1	—	—	—	—	—	—	—	—	—	1	—	1	—	3
Italy	29	—	—	19	19	64	6	12	21	42	40	—	7	4	—	3	—	266
Japan	—	—	—	—	—	—	—	—	—	—	—	—	1	—	—	—	—	1
Netherlands	3	1	—	5	1	—	81	1	97	—	5	—	—	3	—	36	3	236
New Zealand	4	—	—	—	—	—	—	—	—	—	—	—	—	—	—	—	—	4
Poland	—	—	—	—	2	—	3	—	—	—	1	—	—	1	—	—	—	7
Portugal	—	—	—	—	—	—	—	—	—	—	—	—	—	1	—	—	—	1
Rumania	—	—	—	—	—	1	7	4	—	—	—	—	1	—	—	7	—	20
S. Africa	9	—	—	—	—	—	—	10	—	—	13	—	1	—	—	—	—	33
Spain[a]	—	—	—	—	—	1	35	30	—	—	3	—	1	4	—	85	—	159
Sweden	153	25	100	45	30	—	127	—	153	—	139	100	25	29	101	140	32	1,199
U.S.A.	43	—	—	2	—	—	—	8	23	—	75	—	—	—	—	—	—	151
Yugoslavia	—	—	—	—	—	—	—	1	—	—	—	—	1	—	—	1	1	4
Unknown	—	4	—	—	—	—	—	—	—	—	—	—	—	—	—	—	—	4
Sum	344	31	100	92	62	100	269	100	305	43	330	100	39	45	101	288	47	2,396

[a] Including Canary Islands.

broccoli, cherries, peas, and tomatoes only occasionally showed medium or strong inhibition. In imported apples, domestic cucumbers, and domestic lettuce the percentage of samples with medium or strong inhibition was 38, 35, and 36, respectively, and in imported pears as high as 54.

In samples with no or weak inhibition no pesticides have been found by gas chromatographic analyses with the phosphorus detector. In samples with medium inhibition traces of certain pesticides have been found but also peaks of what might be phosphorus compounds, e.g., pesticide metabolites. In most of the samples with strong inhibition one or several pesticides have been identified and quantitated by gas chromatography.

In carrots, cauliflower, potatoes, and white cabbage there are naturally occurring substances giving 30 to 60 percent enzyme inhibition which makes it impossible to detect small amounts of pesticides in the screening test of crude extracts. Cleanup removes these substances to some extent. Figure 2 (Group B) shows the enzyme inhibition of the bromine-treated, purified extracts of these crops. Most of the samples showed no or weak inhibition, 20 percent of the samples medium, and eight percent strong inhibition. Of these crops carrots is the only one where pesticides have been detected by gas chromatography.

c) *Results of gas chromatographic analysis*

Figure 2 (Group A and B) also summarizes the results of gas chromatography. All samples which showed medium or strong cholinesterase inhibition were chromatographed as well as some with no or weak inhibition. The values in brackets give the percentage of the collected samples analysed by gas chromatography. The bars indicate the percentage of samples in which pesticides were actually found. The samples were distributed into three groups with regard to the content of identified pesticides. The amount of residues in these groups are 0.01 to 0.09 mg./kg., 0.10 to 0.50 mg./kg., and >0.50 mg./kg., respectively. The level of 0.01 mg./kg. is the minimum amount of the identified pesticides which can be determined with good accuracy. The level of 0.50 mg./kg. is a common tolerance for organophosphorus pesticide residues in fruits and vegetables in many European countries. When there were several pesticides in the same sample it was grouped according to the total amount of pesticides.

In the whole material, nine different pesticides have been found, identified, and quantitated. Figure 3 gives data on the occurrence of the individual pesticides. They occurred in eight percent of the total number of samples (190 samples out of 2,396). The amounts of the pesticides found have been distributed into the same three groups as in Figure 2. In four percent of the samples the residues were 0.10 mg./kg. or higher. These samples are described in detail in Table VI.

d) *Crops with high residue content*

Apples, pears, peaches, lettuce, and carrots are the crops in which residues have been found most frequently (Fig. 2).

In 191 imported apple samples residues of 0.01 to 0.93 mg./kg. were found in 16 percent. Ethion was found in 8.4 percent and parathion in 5.2 percent. In 153 domestic apple samples residues (parathion 0.01 to 0.03 mg./kg.) were found in 2.6 percent.

In 191 imported pear samples residues of 0.01 to 0.78 mg./kg. were found in 44 percent; ethion was found in 23.6 percent and parathion or parathion-methyl in 16.8 percent. In 139 domestic pear samples residues of 0.01 to 1.75 mg./kg. were found in 3.6 percent.

The peaches contained parathion and parathion-methyl residues of 0.01 to 0.45 mg./kg. in 11.6 percent of the samples. Peaches are not

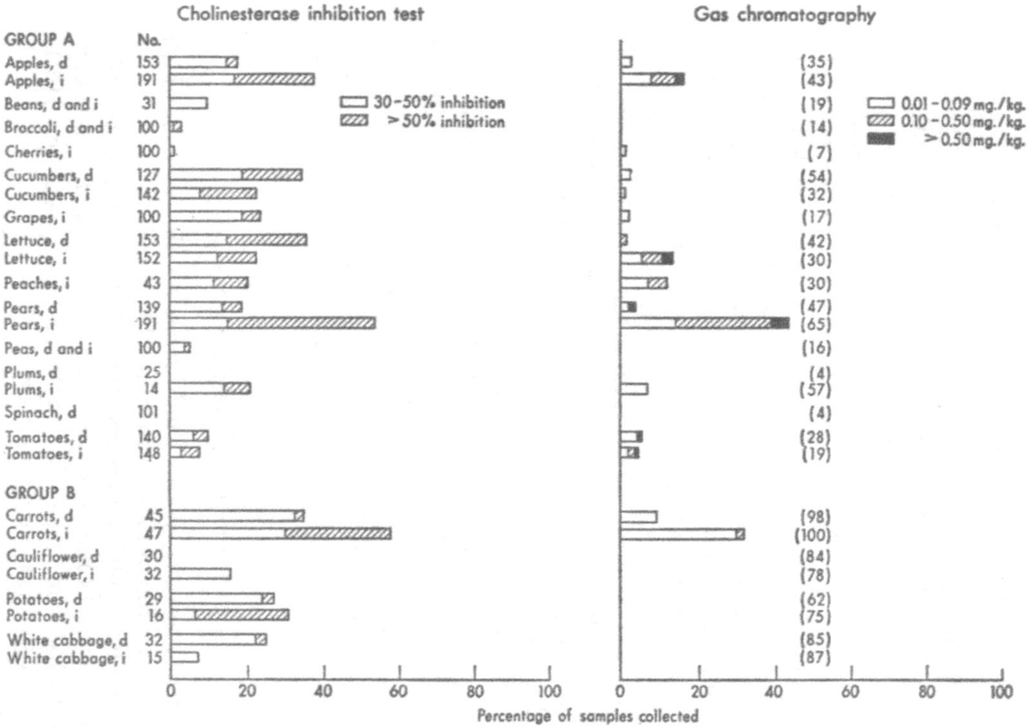

Fig. 2. Occurrence of samples with cholinesterase inhibition or with pesticides determined by gas chromatography. The cholinesterase inhibition test was made on crude extracts of samples in group A and on extracts purified with an alumina column in group B. "No." indicates the number of samples collected. The figures in brackets denote the percentage of samples actually analyzed by gas chromatography. d = domestic, i = imported

grown commercially in Sweden and are sold only during about three months, and thus form only a small part of the total diet.

Of 152 samples of imported lettuce 5.3 percent had 0.01 to 0.09 mg./kg., 5.3 percent 0.10 to 0.50 mg./kg., and 2.6 percent more than 0.50 mg./kg. of pesticide residues. The pesticides were parathion (0.01 to 4.6 mg./kg.) in 16 samples, diazinon (0.08 to 0.78 mg./kg.) in three samples, and parathion-methyl (0.27 mg./kg.) in one sample. Most of the samples with residues were imported from The Netherlands during the months of November to March and were probably greenhouse produce. Two domestic samples out of 153 analysed (1.3 percent) had 0.11 and 0.19 mg./kg. of parathion. It should be noted however, that many of the samples of domestic lettuce showed cholinesterase inhibition, which might be due to unidentified phosphorus pesticides or carbamates.

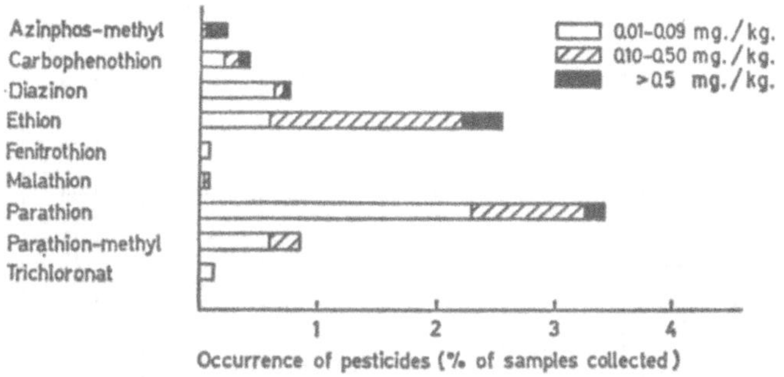

Fig. 3. Frequency of occurrence of the various identified pesticides in the total number of samples

About one-third of the 47 samples of imported carrots contained detectable amounts of pesticides, usually diazinon or parathion. Only one of the samples contained more than 0.1 mg./kg. In the 45 domestic samples residues were found in nine percent and none with more than 0.02 mg./kg.

In the other crops residues have been observed only in zero to five percent of the samples. The value of seven percent for imported plums is uncertain since only 14 samples were analysed. Most of these crops, however, contained residues below the 0.10 mg./kg. level.

There was a big difference in residue content between imported and domestic produce. Thus, residues were found in 13.8 percent of the 1,197 samples of imported produce (of which 7.6 percent $\geqslant$0.10 mg./kg.) compared with 2.1 percent (of which 0.3 percent $\geqslant$0.10 mg./kg.) of the 1,199 domestic samples.

Table VI. *Samples with pesticide residues*[a] *exceeding 0.1 mg./kg. fresh weight, ranged chronologically within each crop species*

Crop and date for arrival	Country	Azinphos-methyl	Carbophe-nothion	Diazinon	Ethion	Parathion	Parathion-methyl
Apples							
65.08.11	France	—	—	—	0.12	—	—
66.01.31	U.S.A.	—	—	—	0.12	—	—
66.02.15	Canada	—	—	—	0.58	—	—
66.03.14	U.S.A.	—	—	—	0.93	—	—
66.03.21	Argentina	—	—	—	—	0.12	—
66.03.21	Argentina	—	—	—	—	0.16	—
66.06.16	Chile	—	0.48	—	—	—	—
67.01.11	U.S.A.	—	—	—	0.37	—	—
67.01.12	U.S.A.	—	—	—	0.22	—	—
67.01.18	U.S.A.	—	—	—	0.31	—	—
67.04.04	Canada	—	—	—	0.50	—	—
68.01.19	U.S.A.	—	—	—	0.87	—	—
68.01.19	U.S.A.	—	—	—	0.14	—	—
68.01.26	U.S.A.	—	—	0.01	0.43	—	—
68.01.26	Canada	—	—	—	0.13	—	—
68.04.26	Chile	—	0.04	—	0.07	—	—
Carrots							
68.02.06	Denmark	—	—	0.27	—	0.23	—
Lettuce							
65.12.02	Netherlands	—	—	—	—	0.16	—
66.01.17	Netherlands	—	—	0.10	—	—	—
66.01.26	Netherlands	—	—	—	—	0.80	—
66.02.14	Netherlands	—	—	—	—	0.83	—
66.06.22	Sweden	—	—	—	—	0.11	—
66.11.24	Netherlands	—	—	—	—	0.16	—
66.12.01	Netherlands	—	—	—	—	4.6	—
66.12.19	Netherlands	—	—	—	—	0.33	—
67.03.02	Netherlands	—	—	—	—	0.33	—
67.10.11	Sweden	—	—	—	—	0.19	—
68.01.17	Netherlands	—	—	0.78	—	—	—
68.11.06	Netherlands	—	—	—	—	0.11	—
68.11.12	Netherlands	—	—	—	—	—	0.27
68.12.04	U.S.A.	—	—	—	—	0.16	—
Peaches							
67.07.10	Italy	—	—	—	—	0.45	0.01
67.08.24	Italy	—	—	—	—	0.26	—
Pears							
64.12.09	U.S.A.	—	—	—	0.23	—	—
65.01.14	U.S.A.	—	—	—	0.12	—	—
65.06.01	Argentina	—	—	—	—	0.12	—
65.12.02	Argentina	—	—	—	0.35	—	—
65.12.06	U.S.A.	—	—	—	0.15	—	—
66.01.10	U.S.A.	—	—	—	0.28	—	—
66.01.31	U.S.A.	—	—	—	0.45	—	—
66.01.31	U.S.A.	—	—	—	0.55	—	—
66.02.15	U.S.A.	—	—	—	0.17	—	—
66.03.08	U.S.A.	—	—	—	0.20	—	—
66.03.08	U.S.A.	—	—	—	0.22	—	—
66.03.08	U.S.A.	—	—	—	0.15	—	—
66.03.08	U.S.A.	—	—	—	0.40	—	—
66.03.08	U.S.A.	—	—	—	0.25	—	—
66.03.08	U.S.A.	—	—	—	0.20	—	—
66.03.08	U.S.A.	—	—	—	0.16	—	—
66.03.14	U.S.A.	—	—	—	0.46	—	—
66.03.14	U.S.A.	—	—	—	0.30	—	—
66.05.17	Australia	—	0.57	—	—	—	—
66.09.29	Sweden	1.75	—	—	—	—	—

Table VI. (continued)

Crop and date for arrival	Country	Azinphos-methyl	Carbophe-nothion	Diazinon	Ethion	Parathion	Parathion-methyl
66.12.01	Italy	—	—	—	—	0.12	0.02
66.12.08	U.S.A.	—	—	—	0.13	—	—
66.12.19	U.S.A.	—	—	—	0.15	—	—
67.01.12	U.S.A.	—	—	—	0.22	—	—
67.01.18	U.S.A.	—	—	—	0.56	—	—
67.01.18	U.S.A.	—	—	—	0.15	—	—
67.01.18	U.S.A.	—	—	—	0.10	—	—
67.02.22	Argentina	—	—	—	—	0.23	—
67.02.22	Argentina	—	—	—	—	0.30	—
67.03.01	Argentina	—	—	—	—	0.11	—
67.03.02	U.S.A.	—	—	—	0.11	—	—
67.03.02	U.S.A.	0.53	—	—	—	—	—
67.03.02	U.S.A.	—	—	—	0.13	—	—
67.03.02	U.S.A.	—	—	—	0.24	—	—
67.03.13	U.S.A.	—	—	—	—	0.18	—
67.03.17	U.S.A.	—	—	—	—	0.25	—
67.03.20	U.S.A.	—	—	—	0.44	—	—
67.03.20	U.S.A.	—	—	—	0.34	—	—
67.04.04	Argentina	—	—	—	—	0.30	—
67.04.04	S. Africa	—	—	—	—	0.10	—
67.05.19	Australia	0.25	—	—	—	—	—
67.05.19	Argentina	—	—	—	—	0.20	—
67.07.17	Italy	0.65	—	—	—	—	—
67.08.07	Italy	—	—	—	—	0.01	0.17
67.08.07	Italy	—	—	—	—	—	0.11
67.08.28	Italy	—	—	—	—	—	0.12
67.10.10	Sweden	0.52	—	—	—	—	—
67.11.23	Italy	—	—	—	—	—	0.20
67.11.24	Italy	—	—	—	—	0.02	0.18
67.12.05	Italy	—	—	0.08	—	—	0.02
67.12.14	U.S.A.	—	—	—	0.78	—	—
67.12.14	U.S.A.	—	—	—	0.75	—	—
67.12.14	U.S.A.	—	—	—	0.70	—	—
68.01.19	U.S.A.	—	—	—	0.11	—	—
68.01.26	Canada	—	—	—	0.40	—	—
68.01.26	U.S.A.	—	—	—	0.20	—	—
68.06.17[b]	S. Africa	—	—	—	—	—	—
68.12.04	U.S.A.	—	—	—	0.11	—	—
Tomatoes							
67.03.02	Spain	—	0.43	—	—	—	—
67.03.20	Spain	—	0.47	—	—	—	—
67.03.13	Spain	—	0.59	—	—	—	—
67.08.24	Sweden	—	—	—	—	0.85	—

[a] The residues were determined by gas chromatography on two different columns and the results were verified by thin-layer chromatography.
[b] Malathion = 0.12 mg./kg.

e) Frequently occurring pesticides

Parathion was the most frequently found organophosphorus pesticide and was found in 3.4 percent of all samples (Fig. 3). It was found mainly in apples, pears, peaches, lettuce, and carrots, but was also found in tomatoes, cucumbers, grapes, and cherries, in imported crops as well as in domestic. Ethion was also often found but only in imported apples and pears. Its residue levels were higher than those of parathion. Carbophenothion occurred in apples, pears, and tomatoes. Like ethion it was found only in imported produce. Ethion and carbophenothion are not registered for use in Sweden. Diazinon and parathion-methyl were found in both fruits and vegetables.

f) Distribution of pesticides between peel and pulp

The distribution of ethion and parathion between peel and pulp was investigated in 12 samples of pears and two samples of apples. The weight of the peel amounted to about ten percent of the whole fruit weight. Eleven of the pear samples contained 0.01 to 0.46 mg. ethion/kg. of whole fruit; 95 to 100 percent (in one sample 90 percent) of the ethion was located in the peel, thus giving 0.07 to 4.4 mg./kg. in the peel and less than 0.01 mg./kg. (in the exceptional sample 0.04 mg./kg.) in the pulp. These samples were taken from the same cargo in March 1966 and were imported from the U.S.A.

One pear sample and the two apple samples contained parathion. The apple samples had about ten percent of the parathion in the pulp and the pear sample had 25 percent in the pulp. This fruit was imported from Argentina and purchased in March 1966.

g) Residues found compared with tolerances in Sweden

The tolerances for pesticides in Sweden were originally prescribed for organochlorine pesticides on March 1, 1966. The values were revised on July 1, 1967. At that time tolerances were also established for ethion and parathion. New tolerances were established on May 15, 1970. The values for fresh fruit, berries, and vegetables are:

Pesticide	*Tolerance (mg./kg.)*	
aldrin plus dieldrin	0.1	(carrots 0.05)
DDT	1	(apples and pears 2)
lindane	1	
ethion	0.5	(apples and pears 1)
parathion	0.5	(apples and pears 1)
diphenyl	70	(only citrus fruit)
o-phenylphenol and its sodium-salt	10	(only citrus fruit, apples and pears)
thiabendazole	3	(only bananas, peeled 0.4)
thiabendazole	6	(only citrus fruit)

No tolerances have as yet (July 1970) been prescribed in Sweden for any other pesticide.

In 19 (three of them domestic) out of the 2,396 samples analysed (0.8 percent) the organophosphorus pesticide level exceeded 0.5 mg./kg. Four of these samples (one domestic) contained parathion in an amount exceeding the Swedish tolerances; the excess was about

50 percent except in one imported sample where the excess was about nine times the tolerance.

WESTÖÖ *et al.* (1969) have analysed about 2,200 food samples collected in Sweden for the content of organochlorine pesticides. They found that 10 samples had a pesticide content exceeding the Swedish tolerances.

h) Methodological aspects

The large number of organophosphorus pesticides in use together with a multitude of toxic metabolites in the crops constitute great problems when crop samples with unknown spray history are analysed. The necessity of a biochemical test method to show the presence or absence of toxicologically significant residues already in the screening step is obvious. The enzymatic methods using cholinesterase are easy to perform in the routine laboratory. Besides a manometric method, the cholinesterase inhibition can be detected on agar sheets (SANDI 1962), on thin-layer chromatograms (MENDOZA *et al.* 1968), or by AutoAnalyzer (LEEGWATER and VAN GEND 1968). However, the cholinesterase method is not sufficient. Some crops may show considerable inhibition even when untreated. Furthermore, potency in cholinesterase inhibition *in vitro* and toxicity to man are not strictly comparable. Therefore the cholinesterase inhibitors must be identified. For identification purposes chromatographic methods are used. The gas chromatographic method used in the present investigation is adequately sensitive and specific for phosphorus compounds.

The manometric method used in this investigation is extremely sensitive. For several of the pesticides an inhibition effect less than 50 percent corresponded to a concentration too low to be detected by gas chromatography in the routine analysis. In samples with more than 50 percent inhibition, however, the pesticides could generally be detected. In crude extracts of the crops listed in Figure 2A 11 percent of the samples had more than 50 percent enzyme inhibition but pesticides were detected chromatographically only in eight percent of the samples. Thus, even in samples with strong enzyme inhibition pesticides could not always be identified. This inhibition might be due to the presence of pesticide metabolites, to pesticides not determinable by the applied method, or to carbamates. It is known that with the present analytical method it is not always possible to detect certain substances, for instance, azinphos-methyl and water-soluble substances such as dimethoate and demeton-S-methyl.

IV. General comments

In this study the chromatographic methods permitted the analysis of 25 organophosphorus pesticides, but no attempt was made to analyse their metabolites. Thus, other phosphorus pesticides or toxic metabolites might be present. Appreciable amounts of them would, however, have been observed in the enzymatic test. In fact, unknown cholinesterase inhibitors were present in cucumbers, lettuce, and potatoes. One might suspect that these unknown inhibitors are derived, at least in part, from pesticides. In addition to organophosphorus substances the carbamate pesticides can also cause inhibition. For the other crop species, the results of the chromatographic and enzymatic analyses agree fairly well, indicating the reliability of the results.

It was found that the domestic produce only rarely contained appreciable pesticide residues whereas in some imported crop species this was frequently the case. The residue levels of the latter crops are probably not representative of the market produce in the exporting country, since crops intended for export must be in extremely good condition, which is partly reached by chemical control of pests and diseases. This obviously may present problems to countries like Sweden with great consumption of imported fruits and vegetables.

In the present investigation a very small fraction of samples exceeded the tolerances recommended by FAO/WHO, which shows that good agricultural practice prevails. But a fairly large fraction contained residues which were only just below the tolerances and an unacceptable exposure to phosphorus pesticides might still occur in the total diet. Samples with such residue levels most often came from countries with a highly industrialized cultivation. It seems probable that systematical pesticide treatment will spread to more countries in the near future. Owing to this and from our data it is recommended that at least some crops, foreign and domestic, be subjected to continuous control. It should also be pointed out that the organophosphorus pesticides have only fairly recently been introduced and it is possible that the tolerances at present recommended should be subjected to revision. The need for international co-operation must be stressed in connection with the control work.

The collection of crop samples of the species reported above is continuing and from 1968 this also includes citrus fruit. The findings obtained so far follow the general trends outlined above.

Acknowledgments

The authors wish to express their special debt to Astrid Carsjö, Ewa Hestad, and Barbro Werelius for their skilful assistance in this work. We also wish to thank the many chemical companies for their generosity in supplying pesticide standards. The financial supports from the Swedish Forest and Agricultural Research Council, the 1964 Committee for the Swedish Natural Resources, and the National Institute of Public Health are gratefully acknowledged.

Summary

Fruit and vegetables produced in or imported into Sweden were analysed for residues of organophosphorus pesticides. The analytical methods, which included cholinesterase inhibition test, gas chromatography, and thin-layer chromatography are described. They permitted the identification of 25 organophosphorus pesticides. From July 1964 to December 1968, 2,396 samples of different produce (apples, beans, broccoli, carrots, cauliflower, cherries, cucumbers, grapes, lettuce, peaches, pears, plums, potatoes, spinach, tomatoes, and white cabbage) were collected. Half of the samples were collected from crops produced in Sweden. Cholinesterase inhibition, which could account for residues of pesticides, occurred in 11 percent of the samples, and in eight percent of these the pesticides were actually identified. In 0.8 percent of the samples (0.7 percent of these were imported and 0.1 percent domestic) there were residues exceeding 0.5 mg./kg., a common tolerance level. A pesticide content of 0.1–0.5 mg./kg. was obtained in 3.2 percent of the samples (3.1 percent imported and 0.1 percent domestic), and 0.01–0.1 mg./kg. was found in four percent (3.0 percent imported and 1.0 percent domestic). The pesticides found were, in descending order of occurrence, parathion, ethion, parathion-methyl, diazinon, carbophenothion, azinphos-methyl, trichloronat, fenitrothion, and malathion. Parathion and ethion, the most frequently occurring, were mainly found in imported apples and pears while parathion was also often found in imported lettuce.

Résumé*

Les résidus de pesticides organophosphorés dans les fruits et les légumes du marché suédois de 1964 á 1968.

Le contrôle de la teneur en résidus de pesticides a été effectué dans des échantillons de fruits et de légumes produits en Suède ou importés. Les auteurs décrivent les méthodes analytiques: inhibition de la cholinesterase, chromatographie gazeuse et en couche mince, méthodes qui permirent l'identification de 25 pesticides organophosphorés. Entre juillet 1964 et décembre 1968, 2396 échantillons de différentes espèces ont été prélevés: cerises, pêches, poires, pommes, prunes, raisins, brocolis, carottes, choux, choux fleurs, concombres, épinards, haricots, laitues, pommes de terre et tomates. La moitié des échantillons provenaient de production suédoises. L'inhibition de la cholinesterase, imputable à des résidus de pesticides, survint pour 11% des échantillons et des pesticides ont été réellement identifiés dans 8% des cas. 0,8% des échantillons (formés de 0,7% de produits

* Traduit par R. Mestres.

importés et 0,1% de produits domestiques) renfermaient des résidus supérieurs à 0,5 mg./kg, tolérance courante. Pour 3,2% des échantillons (3,1% importés et 0,1% domestiques) les teneurs étaient comprises entre 0,1 et 0,5 mg/kg, dans 4% des cas (3% importés et 1% domestiques) elles variaient entre 0,01 et 0,1 mg/kg. Les pesticides trouvés étaient, par ordre d'importance décroissante: parathion, éthion, méthyl parathion, diazinon, carbophénothion, azinphos méthyl, trichloronate, fénitrothion et malathion. Le parathion et l'éthion, pesticides observés le plus fréquemment, ont surtout été decelés dans les pommes et les poires d'importation. Des résidus de parathion étaient également fréquents dans les laitues importées.

Zusammenfassung*

Bestimmung der Rückstände organischer Phosphor-Pestizide in Obst und Gemüsen auf den Schwedischen Märkten in den Jahren 1964 bis 1968

In Schweden produziertes oder nach Schweden importiertes Obst und Gemüse wurden auf Rückstände organischer Phosphor-Pestizide analysiert. Die analytischen Methoden, umfassend Cholinesterase-Hemm-Test, Gas- und Dünnschicht-Chromatographie, werden beschrieben. Sie gestatteten die Identifikation von 25 verschiedenen organischen Phosphor-Pestiziden.

In der Zeit von Juli 1964 bis Dezember 1968 wurden insgesamt 2396 Proben von verschiedenen Produkten (Äpfel, Bohnen, Broccoli, Möhren, Blumenkohl, Kirschen, Gurken, Weintrauben, Salat, Pfirsiche, Birnen, Pflaumen, Kartoffel, Spinat, Tomaten und Weisskohl) genommen. Die Hälfte der gesammelten Proben stammten von Produkten schwedischer Herkunft.

Cholinesterase-Hemmung, die als Hinweis für die Anwesenheit von Pestizid-Rückständen gilt, wurde in 11% der Proben gefunden: in 8% konnten die Pestizide auch identifiziert werden. Bei 0,8% der Proben (0,7% umfassten importierte, 0,1% einheimische Produkte) waren die Rückstände höher als 0,5 mg/kg und überschritten damit die übliche Toleranz-Grenze. Ein Pestizid-Gehalt von 0,1 bis 0,5 mg/kg wurde in 3,2% der Proben (davon 3,1% importierte und 0,1% einheimische Produkte), ein Gehalt von 0,01 bis 0,1 mg/kg wurden in 4% der Proben (davon 3,0% importierte und 1,0% einheimische Produkte) gefunden.

Die nachgewiesenen Pestiziden waren, in der Reihefolge ihrer Häufigkeit folgende: Parathion, Ethion, Methyl-Parathion, Diazinon, Carbofenothion, Methyl-Azinfos, Trichloronat, Fenitrothion und Malathion. Parathion und Ethion, die am Häufigsten gefundenen Pestizide, wurden hauptsächlich in importierten Äpfeln und Birnen, aber Parathion auch in Salat, gefunden.

---

* Übersetzt von H. F. LINSKENS.

References

ABBOTT, D. C., and H. EGAN: Determination of residues of organophosphorus pesticides in foods. Analyst **92**, 475 (1967).

———, D. C. HOLMES, and J. O'G. TATTON: Pesticide residues in the total diet in England and Wales, 1966–1967. II. Organochlorine pesticide residues in the total diet. J. Sci. Food Agr. **20**, 245 (1969).

ARCHER, T. E.: Enzymatic methods. In G. Zweig (ed.): Analytical methods for pesticides, plant growth regulators, and food additives, Vol. I, p. 373. New York: Academic Press (1963).

AUGUSTINSSON, K. B.: Assay methods for cholinesterases. Meth. Biochem. Anal. **5**, 1 (1957).

COFFIN, D. E.: Residues of parathion, methyl parathion, EPN, and their oxons in Canadian fruits and vegetables. Residue Reviews **7**, 61 (1964).

DUGGAN, R. E., and J. R. WEATHERWAX: Dietary intake of pesticide chemicals. Science **157**, 1006 (1967).

FALLSCHEER, H. O., and J. W. COOK: Studies on the conversion of some thionophosphates and a dithiophosphate to *in vitro* cholinesterase inhibitors. J. Assoc. Official Agr. Chemists **39**, 691 (1956).

FAO/WHO: 1967 evaluations of some pesticide residues in food. FAO/PL: 1967/M/11/1; WHO/Food Add./68.30 (1968a).

——— Pesticide residues; report of the 1967 joint meeting of the FAO working party of experts and the WHO expert committee. FAO Meeting Report No. PL: 1967/M/11; World Health Org. Tech. Rept. Ser., No. 391 (1968b).

——— 1968 evaluations of some pesticide residues in food. FAO/PL: 1968/M/9/1; WHO/Food Add./69.35 (1969a).

——— Pesticide residues in food; report of the 1968 joint meeting of the FAO working party of experts on pesticide residues and the WHO expert committee on pesticide residues. FAO Agricultural Studies No. 78; World Health Org. Tech. Rept. Ser., No. 417 (1969b).

GAGE, J. C.: Residue determination by cholinesterase inhibition analysis. Adv. Pest Control Research **4**, 183 (1961).

HARRIES, J. M., C. M. JONES, and J. O'G. TATTON: Pesticide residues in the total diet in England and Wales, 1966–1967. I. Organisation of a total diet study. J. Sci. Food Agr. **20**, 242 (1969).

HARTMANN, C. H.: Phosphorus detector for pesticide analysis. Bull. Environ. Contamination Toxicol. **1**, 159 (1966).

ISO (International Organization for Standardization): Common names for pesticides. Lists 1–14 (1959–1969).

JONES, K. H., D. M. SANDERSON, and D. N. NOAKES: Acute toxicity data for pesticides (1968). World Review Pest Control **7**, 135 (1968).

KOVACS, M. F., JR.: Thin-layer chromatography for organothiophosphate pesticide residue determination. J. Assoc. Official Agr. Chemists **47**, 1097 (1964).

KRAUSE, C., and J. KIRCHHOFF: Organophosphatrückstände auf Marktproben von Obst und Gemüse sowie auf Getreideerzeugnissen. Nachrichtenbl. deutsch. Pflanzenschutzd. **21**, 81 (1969).

LAWS, E. Q., and D. J. WEBLEY: The determination of organo-phosphorus insecticides in vegetables. Analyst **86**, 249 (1961).

LEEGWATER, D. C., and H. W. VAN GEND: Automated differential screening method for organophosphorus pesticides. J. Sci. Food Agr. **19**, 513 (1968).

MENDOZA, C. E., P. J. WALES, H. A. MCLEOD, and W. P. MCKINLEY: Thin-layer chromatographic-enzyme inhibition procedure to screen for organophosphorus pesticides in plant extracts without elaborate clean-up. Analyst **93**, 173 (1968).

MITCHELL, T. H., J. H. RUZICKA, J. THOMSON, and B. B. WHEALS: The chromatographic determination of organophosphorus pesticides. Part III. The effect of irradiation on the parent compounds. J. Chromatog. **32**, 17 (1968).

SÁNDI, E.: Neue Wege zum Nachweis und zur Bestimmung der Rückstände insektizider Phosphorsäureester in Lebensmitteln. Die Nahrung **VI**, 57 (1962).

THOMPSON, R. H., and E. G. HILL: Pesticide residues in foodstuffs in Great Britain. XI. Further studies of malathion in imported cereals. J. Sci. Food Agr. **20**, 293 (1969).

WATTS, R. R.: 4-(*p*-Nitrobenzyl) pyridine, a new chromogenic spray reagent for the organophosphate pesticides. J. Assoc. Official Agr. Chemists **48**, 1161 (1965).

WESTÖÖ, G., K. NORÉN, and M. ANDERSSON: Rester av vissa klorpesticider och arsenik i frukt, bär, grönsaker och rotfrukter i svensk handel åren 1964–1967. Vår föda, nr **9**, p. 73 (1968).

——— ——— ——— Analyses of some organochlorine pesticides and arsenic in fruits, berries, and vegetables on the Swedish market in 1964–1967. Acta Agr. Scand. **19**, 33 (1969).

Manuscript received 22 December 1969; accepted 6 February 1970.

Pesticide residue analysis in the presence of Polychlorobiphenyls (PCB's)

By

L. M. Reynolds*

Contents

I. Introduction

Although there has been so much discussion on the pros and cons of pesticide use, it would be unthinkable to start this review without a few comments on the controversy. The recent banning of DDT by some countries and the widespread and increasing animosity towards the use of the organochlorine pesticides in general are due to a number of factors including:

a) The large annual additions of the pesticides to the environment coupled with their apparent persistence.
b) Their known acute toxicities.
c) Their widespread distribution in biological materials and their reported buildup in the food chains.

* Ontario Research Foundation, Sheridan Park, Ontario, Canada.

d) Their more recent implication in the derangement of calcium metabolism (PEAKALL 1967, SIMKIS 1967, WELCH *et al.* 1969) possibly causing the decrease in eggshell thickness and the population decline of certain raptorial species of birds (RATCLIFFE 1967, HICKEY and ANDERSON 1968, STICKEL 1969).
e) Their adverse effects on fish reproduction (BURDICK *et al.* 1964) and high residue levels found in Coho salmon in Lake Michigan in 1969.
f) Certain "fears" involving chronic toxicities and sublethal doses with possible deleterious effects on man and wildlife and the ultimate survival of man.

The above factors are very compelling and certainly warrant man's concern about the threat to his environment and his very existence. Nevertheless, one cannot deny that these pesticides have been important in ensuring the supply of man's rapidly increasing food requirement and in the protection of his health, as well as the saving of millions of lives. Surely, this is a case where the benefit-risks equation as well as the alternatives to the organochlorine pesticides should be carefully studied and evaluated to avoid the possibility of getting into a worse predicament.

It was quite ironic when recently a large city publicised that, because of an impending ban on DDT, the garbage collector would pick up separately all home garden sprays and other materials which contained even traces of DDT. The material that had been used more than generously and cherished for almost 30 years had suddenly become monstrous even in trace quantities. Wasn't there much greater danger posed to the collectors, children, and animals by such collection than if the home supplies had been allowed to run out gradually with no chance of replenishment?

In fact, the announced bans on DDT so far are generally not total, and hence serve the same purpose as merely tightening the control on the uses of such pesticides. In other words, when it is necessary, and until suitable and satisfactory replacements are available, limited and judicious use will be made of these pesticides.

With regard to the use of substitutes, caution is essential to avoid premature use of any compounds. It is now known that potentiation (or antagonism) of toxicity is possible with certain combinations of pesticides and with combinations of pesticide and drugs or environmental chemicals (DUBOIS 1968). A compound might be fairly safe by itself yet harmful in the presence of others.

It is a fact that even if suitable substitutes were found and the use of the organochlorine pesticides was banned completely, a need would still exist for at least a number of years to have reliable analytical methods for the determination of their residues. The present review discusses the status of organochlorine pesticide residue analysis with particular reference to serious inaccuracies caused by the presence of polychlorobiphenyls (PCB's) in some samples.

As would be expected, in order to cope with the numerous new pesticides that have been put into use, pesticide residue methodology has changed drastically over the years. ZWEIG (1963) pointed out that until about 1940, the life of an analytical chemist working in the pesticide field was a relatively serene one in which the chemist had to be familiar only with analytical methods for arsenic, lead, fluoride, pyrethrins, rotenone, and a few others. Just about ten years back, DDT and other organochlorine compounds were estimated by colorimetric methods (SCHECTER *et al.* 1945) which obviously did not give precise information on the amounts and identities of the individual compounds present.

Today's residue methodology is much more complex, and in order to cope with the new pesticides being introduced, the techniques are constantly being modified and improved. The wide variations in toxicity and persistence of pesticides, new or old, dictate refinements in methods in order to differentiate and determine individual residues.

Bearing in mind the complexities mentioned above and the fact that the spray history of most samples (excluding controlled spraying experiments) is unknown to the analyst, the availability of chromatographic techniques which afford separation of a number of compounds in a mixture has made the multi-residue screening approach[1] (*U. S. Food and Drug Administration* 1969 revision) the method of choice.

The gas liquid chromatographic (GLC) separation with electron capture (EC) detection is presently the most popular analytical combination for organochlorine pesticide residues.

Multi-residue analysis involving GLC generally involves five principal steps:

a) Sampling to procure a representative aliquot of the sample.
b) Extraction of the residue with a solvent (usually organic since most of the organochlorines (OC's) and many of the organophosphates (OP's) are fat soluble).
c) Cleanup of the extract to remove interfering materials such as fats, waxes, pigments, etc.
d) Analysis of the cleaned-up extract, generally by GLC-EC technique.
e) Confirmation of the pesticide identity.

Certainly there are a great number of variations in carrying out these steps, depending on the nature of the pesticide and its substrate,

[1] Editor's note: See BURKE, J. A.: Development of the Food and Drug Administration's method of analysis for multiple residues of organochlorine pesticides in foods and feeds. Residue Reviews, this volume.

the limit of detectability required, and the laboratory staff and equipment available. It is beyond the scope of this review to discuss any of these methods in detail. Furthermore, Gunther (1962) and Samuel and Hodges (1967) have reviewed this topic adequately. Accordingly, this review will deal mainly with problems involving PCB's and related compounds. No attempt will be made to review gas chromatography for pesticide residue analysis. Burke (1965) has discussed some practical aspects of this topic, while Thompson *et al.* (1969 and 1969 a) have evaluated different gas chromatographic columns for chlorinated pesticides.

Before discussing the PCB's and their interference with pesticide residue analysis, some[2] key references in the English language on residue methodology are appropriate. These are listed alphabetically as follows:

1. Advances in pest control research, vols. 1–8. New York: Interscience (1957 *et seq.*)
2. A guide to the analysis of pesticides by gas chomatography, 2nd ed. S. T. Preston, Jr. Evanston, Ill.: Polyscience (1968).
3. Analysis of insecticides and acaricides, Gunther and Blinn. New York-London: Interscience (1955).
4. Analytical methods for pesticides, plant growth regulators and food additives, vols. I–V, Zweig. New York: Academic Press (1963).
5. Guide to the analysis of pesticide residues, vols. 1 and 2, Burchfield and Johnson. *U. S. Public Health Service,* Office of Pesticides, Washington, D. C. (1965).
6. Manual of methods for the determination of residues of Shell pesticides, *Shell Chemical Company,* Agricultural Chemicals Division, New York (1969).
7. Official methods of analysis, *Association of Official Analytical Chemists,* 10th ed. Washington, D. C. (1965).
8. Pesticide analytical manual, vols. I and II, *U. S. Department of Health, Education, and Welfare, Food and Drug Administration,* Washington, D. C. (1968).
9. Pesticide residue analysis handbook, Bonelli. *Wilkins Instrument and Research, Inc.,* Walnut Creek, Calif. (1965).
10. Residue Reviews, vols. 1–34. Gunther. New York: Springer-Verlag (1962 *et seq.*)

It should be emphasized that these references were all written prior to the knowledge that there was a possibility of PCB interference with

[2] Editor's note: There are now more than 95 books concerned with pesticide residues including 60 that discuss analytical methodology in some respects; eight countries are represented by the authors. Gunther, F. A.: Pesticide residues in the total environment-Reliable detection and determination, mitigation, and legislative control and surveillance programs. Pure and Applied Chem. **21**, 355 (1970).

certain pesticide residue analyses. In general, the methods described in these references can be modified to cope with PCB interference.

II. Presence of PCB's in the environment

Following the development of the EC detectors and their use with GLC, most residue chemists have observed unidentified peaks on their gas chromatograms. Generally, there was no specific pattern to these peaks, but analysts concerned with pesticide residues in wildlife, especially in Europe (ROBURN 1965, HARRISON 1966, HOLDEN and MARSDEN 1966), sometimes observed a particular series (with as many as ten or more peaks) of the unidentified peaks in their sample extracts. The materials giving rise to these unidentified responses occurred most frequently and in the largest proportions in extracts from aquatic raptorial species of birds and fishes. The general supposition was that these responses were from condensation products of the metabolites of the organochlorine pesticides, since they were generally observed when large amounts of pesticides were also present. ROBURN (1965) reported them to be organochlorine compounds with fairly high chlorine contents, which no doubt accounts for their high sensitivities to the EC detector. However, JENSEN (1966) was the first to state that these unidentified peaks corresponded to PCB compounds, based on the analysis of a number of pike samples, an eagle, and human hair. Initially, the GLC and thin-layer chromatographic (TLC) patterns and chemical inertness were used to identify the unidentified peaks as PCB's (JENSEN and WIDMARK 1967).

Other workers in Great Britain (HOLMES *et al.* 1967, HOLDEN and MARSDEN 1967) and the Netherlands (KOEMAN *et al.* 1967) also made early reports on the presence of PCB's in their fish and wildlife samples.

At the Organisation for Economic Co-operation and Development (OECD) Conference on "The Unintended Occurrence of Pesticides in the Environment," held in Scotland in September, 1967, the European chemists in particular showed much concern regarding the presence of PCB's in their wildlife samples and the possibility of interference with their organochlorine pesticide residue analysis. Up to that time, there had been no report on the occurrence of these materials in North American samples. However, in a December, 1967, report on organochlorine pesticides in seals and porpoises, HOLDEN and MARSDEN included a comparison of the amounts of PCB-type materials found in Scottish and Canadian samples. They reported that the levels of these non-hydrolysable materials were generally much lower in the Canadian specimens. Since that time RISEBROUGH *et al.* (1968) have reported that the PCB's are widely dispersed in the global ecosystem, while REYNOLDS (1968, 1969, and 1969 a) has found them in some Canadian wildlife samples.

Positive confirmation of the PCB's by gas chromatography-mass spectrometry (GC-MS) has now been reported in Sweden (WIDMARK 1967), the Netherlands (KOEMAN *et al.* 1969), and in the United States (REICHEL 1969).

III. PCB interference with pesticide residue analysis

The polychlorobiphenyls (× indicates the possible chlorine positions) and related compounds (polychlorinated triphenyls, naphthalenes, terpenes, etc.) have numerous important industrial uses but are not used as pesticides. Because of their similarities in structure and properties to the DDT pesticide group, the PCB's—if present in a substrate—are carried through the usual pesticide extraction and screening procedures, and since they possess electron absorbing properties, will interfere with GLC-EC analysis of organochlorine as well as a number of the organophosphorus pesticides (REYNOLDS 1969 b).

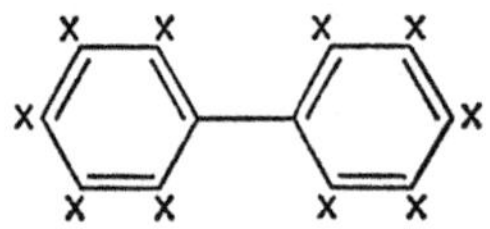

The reported GLC patterns indicating PCB interferences have shown marked resemblances with 'Aroclor' 1254 and 1260 (JENSEN and WIDMARK 1967, HOLMES *et al.* 1967, REYNOLDS 1968 and 1969, and KOEMAN *et al.* 1969). This type of interference with pesticide residue analysis is demonstrated with a standard mixture of organochlorine pesticides and a commercial PCB mixture (Aroclor 1254) in the chromatograms of Figure 1. The results confirm that the presence of PCB's in the sample extracts will cause interference with pesticide residue analysis under these or similar operating conditions.

It should be emphasized that throughout this review when the term "PCB" is used, this includes the polychlorobiphenyls and also the other closely related compounds mentioned above. However, the polychlorobiphenyls are the most important from the point of view of interference with pesticide residue analysis. Under the GLC operating conditions, the chlorinated polyphenyls and other related compounds with molecular weights greater than the chlorinated biphenyls will generally not be separated and observed on the gas chromatograms. Modification of the operating parameters such as higher column temperatures or changes in the liquid stationary phases are usually necessary to detect the presence of such compounds.

It must be borne in mind that when high molecular weight compounds are present, their peaks can appear on subsequent chromatograms and cause "secondary interference." This occurrence can be recognised easily by the early emergence of unusually broad interfering peaks on the chromatograms. "Secondary interference" is a term used by the author to imply that the interfering peak does not have the same retention time as that of the compound of interest under the same operating conditions. The interfering peak often originates from a previous injection and cannot be simply repeated. Of course, if the elution time is very long, the peak might be broad enough to escape detection and in such a case would not interfere. In contrast

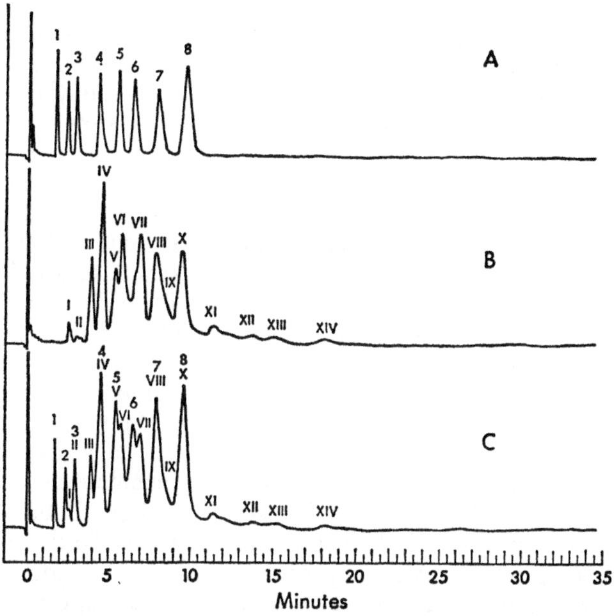

Fig. 1. PCB interference with organochlorine pesticide residue analysis on four percent SE–30/six percent QF–1 on 60/80 mesh Chromosorb W, ⅛″ × 6′ borosilicate column. Chromatogram *A:* standard mixture of organochlorine pesticides; peak numbers: 1 = 0.08 ng. of lindane, 2 = 0.10 ng. of heptachlor, 3 = 0.10 ng. of aldrin, 4 = 0.14 ng. of heptachlor epoxide, 5 = 0.20 ng. of DDE, 6 = 0.20 ng. of dieldrin, 7 = 0.30 ng. of DDD, and 8 = 0.50 ng. of DDT. Chromatogram *B:* 5 ng. of Aroclor 1254 (the 14 major peaks are numbered I to XIV). Chromatogram *C:* combination of above organochlorine standard pesticide mixture and Aroclor 1254. Injector 250°C., column 200°C., detector base 250°C., N_2 at 20–30 cc./minute

to this type of interference, materials which have identical retention times will cause "primary interference" which is easily repeated by reinjection of the extract. With this form of interference a single, sharp peak is obtained, and in the absence of confirmatory and other precautionary measures, misinterpretation can occur easily. PCB-type interference falls mainly in this group.

IV. Uses and properties on PCB's

In order to understand how it is possible for PCB's to be present in wildlife and cause interference with pesticide residue analysis, it is necessary to give a brief description of some of their properties and uses. The PCB's are produced and marketed under a number of commercial trade names, e.g., 'Aroclor,' 'Clophen,' and 'Phenochlor.' Koeman *et al.* (1969) reported marked resemblance between Aroclor 1260, Clophen A60, and Phenochlor DP6. The 'Aroclor' plasticizers, a series of chlorinated biphenyls and chlorinated polyphenyls, are typical of these compounds. As stated by the manufacturers (*Monsanto Co.* 1967), the 'Aroclor' compounds are among the most versatile chemically produced materials available. They vary from mobile, oily liquids to white crystals and hard transparent resins. They are non-oxidizing, inert, permanently thermoplastic, of low volatility, noncorrosive to metals, insoluble in water, and resistant to alkalies, acids, and corrosive chemicals. The viscous, more highly chlorinated liquid and resin members do not support combustion and they impart fire retardance to other materials. The crystalline 'Aroclor' compounds are relatively insoluble but the liquid and resinous compounds are soluble in most of the common organic solvents, thinners, and oils.

These compounds are used in protective coatings, as placticizers and extenders, as sealers in water-proofing compounds and putty, in asphaltic materials, printing inks, waxes, and synthetic adhesives. Liquid PCB's are used as dielectrics, as hydraulic fluids, in thermostats, in cutting oils, as extreme pressure lubricants, as grinding fluids, and as heat transfer media. Solid PCB's are used to impregnate carbon resistors and as sealers in impregnating agents for electrical apparatus. The properties of many products can be varied and improved by the use of these compounds either as primary or secondary plasticizers.

Further details concerning the different 'Aroclor' series, the gradation of their properties, and the numerous possible applications can be found in the manufacturer's technical bulletin (*Monsanto Co.* 1967). In the system of designation for the 'Aroclors,' the first two digits indicate the type of material, while the last two digits give the approximate weight percentage of chlorine in the product. For example, 'Aroclor' 1254 indicates a chlorinated biphenyl with approximately 54 percent chlorine, while 'Aroclor' 5460 indicates a chlorinated triphenyl containing about 60 percent chlorine.

V. Possible modes of entry of PCB's into the ecosystem

The means by which PCB's enter the ecosystem to contaminate fish and wildlife are still not clearly understood. There are however, four most likely pathways by which PCB's enter fish and wildlife:

a) Because of their inertness and versatility and consequent numerous applications, it is quite conceivable that PCB's could be flushed as wastes into rivers, lakes, etc. to pollute fish and other wildlife.
b) There is a possibility that some contamination could proceed via the atmosphere when wastes containing these compounds are burnt.
c) Although it is unlikely, some type of Ullman reaction with condensation of aromatic halides with the aid of metallic agents such as copper to give rise to the formation of biaryls, cannot be completely ruled out.
d) Lastly, but not necessarily the least likely, there is the possibility that some companies are using PCB's in some insecticide formulations to increase the kill-life.

The idea of increasing the residual persistence of insecticides has been investigated for some time and a brief résumé on the more important background work is in order.

The early investigations by LINDQUIST *et al.* (1945) and BLOCK (1948) indicated that the residual effectiveness of DDT was prolonged when the DDT was applied in resins and paints. VAN TIEL (1952) continued along these lines when he increased both the effective period and the toxicity of a DDT deposit on a glass surface by the addition of a small amount of coumarone resin to a DDT solution in kerosene. Subsequently, emphasis was placed on work involving the more volatile insecticides like lindane—a very effective pesticide but one which soon loses its activity because of volatilization.

SULLIVAN and HORNSTEIN (1953) in their experiments with the American cockroach, *Periplaneta americana,* found that lindane residue remained toxic longer when it was compounded with 'Aroclor' 5460. They suggested that the chlorinated polyphenyl prevents crystallization of the lindane and lowers the vapor pressure of the lindane in the mixture, thereby preventing unsightly residues on surfaces as well as extending the effective kill-life of the insecticide. TSAO *et al.* (1953) obtained similar results against house flies, *Musca domestica* L. Continuing this work, HORNSTEIN and SULLIVAN (1953) confirmed these findings and described a general method for prolonging the residual effectiveness of volatile insecticides. The method consisted of preparing concentrated solutions of the pesticides in the film-forming polychlorinated polyphenyls. A *Monsanto Co.* Technical Bulletin (1965) states that because "'Aroclors' in formulations 'trap' and hold more volatile ingredients, they make volatile insecticides and repellents 'last longer' in residual activity."

Attempts to determine whether this idea had been put into practice have so far been unsuccessful. However, if the PCB's are being used in pesticide formulation, then this coupled with the numerous other uses suggested in a later *Monsanto Co.* Bulletin (1967), might certainly explain their presence in wildlife tissues and other samples.

There is an obvious need for clarification of the sources of PCB's in fish and wildlife, and this could probably be accomplished using tracer techniques.

VI. Toxicity of PCB's

The PCB's were studied as early as 1881 (SCHMIDT and SCHULTZ) and were in wide use by 1930 (PENNING 1930). Later JONES and ALDEN (1936) reported that the compounds were toxic. GREENBERG *et al.* (1939) reported that PCB's and polychlorinated naphthalenes caused the deaths of three workers. Men employed in the production and use of PCB's developed acne-type skin eruptions (SCHWARTZ and BARLOW 1942, SCHWARTZ and PECK 1943, SCHWARTZ 1943).

MILLER (1944) reported that PCB's caused pathological changes in laboratory animals. BROWN (1947) warned about the toxicity dangers of the 'Aroclors' when there was a suggestion that one of the 'Aroclors' be used as the melting point bath liquid in preference to the customary sulfuric acid. Later, MCLAUGHLIN *et al.* (1963) found that Aroclor 1242, besides being toxic, produced teratogenic effects in chick embryos at lower dosage levels. RISEBROUGH *et al.* (1968) reiterated that PCB's along with other chlorinated biocides, such as DDT, could account for a large part of the aberration in calcium metabolism that has been observed in many species since the second world war. With high dosage of PCB's, hydropericards have been observed in quails (KOEMAN *et al.* 1969).

It has been stated (*Monsanto Co.* 1967) that at ordinary temperatures the 'Aroclor' chlorinated polyphenyls do not present industrial toxicological problems. However, caution was advised in handling these materials. The hazard of potential toxic exposure varies with the volatility of the compounds: the lower chlorinated, more volatile ones present more of a potential problem from the standpoint of both inhalation and skin contact. At elevated temperatures, the use of PCB's requires very effective and efficient exhaust ventilation systems.

It has also been stated by the manufacturers that "tests on animals indicate that the maximum safe concentration of vapor is in the range of from 0.5 to 1.0 milligram of the lower chlorinated 'Aroclor' plasticizers per cubic meter of air. The threshold limits (maximum allowable concentration for an 8-hour working day) set by the American Conference of Government Hygienists are 1.0 milligram of the lower chlorinated 'Aroclor' compounds per cubic meter of air and 0.5 milligram of the more highly chlorinated compounds, such as 'Aroclor' 1254, per cubic meter of air."

However, as RISEBROUGH *et al.* (1968) pointed out, only relatively small amounts of chlorinated hydrocarbons are required to cause enzyme-induced breakdown of steroids, thus making irrelevant much of the p.p.m. approach to pollutant ecology based on toxicity data alone.

There is no doubt that further research in this area is urgently needed.

VII. Analysis of pesticide residues in the presence of PCB's

PCB's are of interest to analysts for three main reasons:

a) The compounds are toxic.
b) They are widely distributed and their uses are increasing.
c) They interfere with GLC determinations of organochlorine and organophosphorus pesticide residues.

The first two reasons have been dealt with earlier and therefore need no further elaboration. Accordingly most of the remainder of this review will be concerned with the analysis of pesticide residues in the presence of PCB's. Also the present methods of estimating the amounts of PCB's will be discussed.

As mentioned earlier, it was emphasized at the 1967 OECD conference that PCB's were in fact being detected in fish and wildlife, especially in Europe. The conference among other things served to alert other chemists and scientists to the problems of PCB's and their interference with pesticide residue analysis. Since the discovery that PCB's are contaminating fish and wildlife and that they interfere with organochlorine pesticide analysis, very little has been published on how to cope with these interferences and how to differentiate the PCB's from pesticides. Based on preliminary work, JENSEN and WIDMARK (1967) suggested a nitration technique based on the inertness of the PCB's compared to the relatively easier nitration of the pesticides. RISEBROUGH *et al.* (1968) reported the use of this method, but REYNOLDS (1969 b) pointed out that nitration is not a suitable approach since some pesticides (lindane, toxapene, Strobane, etc.) apparently will not nitrate, while some of the PCB's appeared to be nitrated under the suggested conditions. In addition, there is the problem of further complication from the interference of the nitroderivatives formed, especially when the DDT-group pesticides are present in large amounts.

To avoid the above complications and at the same time facilitate the chromatographic interpretations and afford a means of estimating the PCB's and pesticides independently, REYNOLDS (1969 b) introduced a Florisil separation scheme. KOEMAN *et al.* (1969) have also reported a similar Florisil column technique to separate the PCB's from the pesticides.

In Reynolds' method, the partially cleaned-up sample extract containing the pesticides and PCB's is placed on a Florisil column and the PCB's are eluted first with *n*-hexane, followed by elution of the pesticides with an ether-hexane mixture. Originally, this technique was intended to serve as the final cleanup step, as well as effecting separation of the PCB's from the pesticides. As will be explained, the method has since been slightly modified (miniaturized) to serve mainly as a separation technique. Of the many pesticides investigated, only heptachlor, aldrin, and the DDT metabolite *p,p'*-DDE are eluted with the PCB's. With the exception of *p,p'*-DDE, these do not present difficult problems since chemical reactions (epoxidation, hydrobromination, etc.) can be used to differentiate heptachlor and aldrin from the PCB's. Furthermore, these two pesticides are more frequently found as the more stable epoxy products which are separable from the PCB's by this scheme.

p,p'-DDE on the other hand presents a more difficult problem on account of its chemical inertness. Its greater stability resembles that of the PCB's, and the compound is not amenable to any simple and convenient method of structure modification to give a product which is readily detected within the usual GLC-EC operating parameters.

Of course the ideal situation would be a convenient addition of one mole of hydrochloric acid to the *p,p'*-DDE to form *p,p'*-DDT while the PCB's are not affected. This would afford confirmation and quantitative estimation of *p,p'*-DDE in the presence of PCB's. Generally under such circumstances it would not be necessary to do an additional Florisil separation. The amount of *p,p'*-DDE present could be estimated indirectly from the *p,p'*-DDT produced since, as will be discussed later, the *p,p'*-DDT (and *p,p'*-DDD) can be separated from the PCB's by the use of highly polar liquid phases.

It should be emphasized that although *p,p'*-DDD and *p,p'*-DDT are mostly separated by the polar phase columns, without introducing a Florisil separation the earlier emerging pesticides are superimposed on the PCB's and hence a separation is necessary to identify and to quantify these latter pesticides.

After the Florisil separations of PCB's and pesticides are made, the two eluates are chromatographed separately on the SE-30/QF-1 column and compared with appropriate standards for quantitation. This is normally followed by confirmation of the identities of the pesticides.

VIII. Confirmation of pesticide identity

Despite the separation of the PCB's and their elimination as sources of interference, the non-specific nature of the EC detector makes it necessary to confirm the identity of the pesticides by additional techniques.

The sequence usually followed by our laboratory is:

a) Injection of the pesticide eluate onto a more polar liquid phase column (e.g., polyester such as DEGA or DEGS). Typical

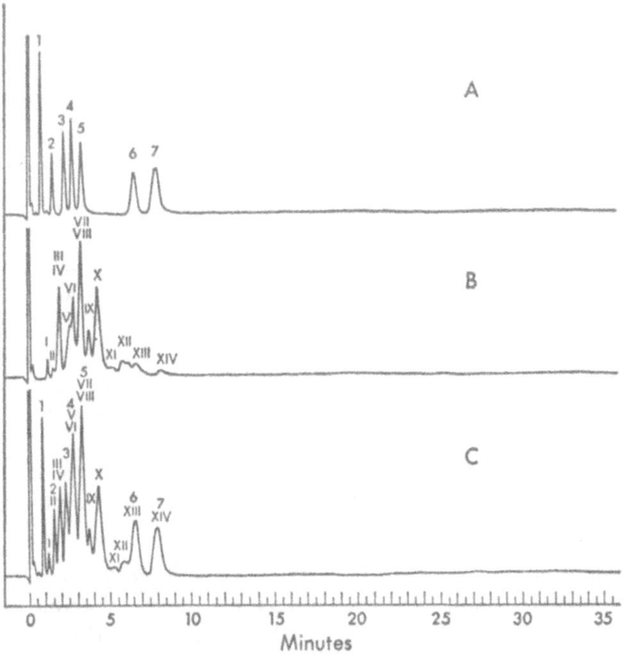

Fig. 2. Separation of PCB's and organochlorine pesticides on polar phase column (five percent DEGS/two percent H_3PO_4). Chromatogram *A:* standard mixture of organochlorine pesticides with same amounts as in Figure 1; peak numbers: 1 = heptachlor and aldrin, 2 = lindane, 3 = heptachlor epoxide, 4 = DDE, 5 = dieldrin, 6 = DDT, 7 = DDD. Chromatogram *B:* 5 ng. of Aroclor 1254 (the 14 major peaks are numbered I to XIV as in chromatogram *B* of Figure 1, assuming that the order of elution is unchanged). Chromatogram *C:* combination of organochlorine standard pesticide mixture and Aroclor 1254. Other GLC parameters as in Figure 1

separations of pesticides, PCB's, and a mixture of the two groups are shown in Figure 2. The reversal of *p,p'*-DDD and *p,p'*-DDT and their separation from most of the PCB's is noteworthy and quite useful. One drawback is that peak No. 13 of Aroclor 1260 interferes partially with *p,p'*-DDT. Other useful information can be obtained by study of the retention times of individual pesticides. For example, this approach affords separation of the three main BHC isomers. The technique is especially useful for ruling out the presence of pesticides indicated as possibly present by the SE-30/QF-1 column. However, the retention times on two or more stationary phases cannot be regarded as independent parameters of identity (Robinson 1967).

b) Derivatization and use of characteristic GLC retention times of the derivatives. As reported earlier (Reynolds 1969 a), our laboratory has increasingly utilized chemical reactions for confirmations of pesticide residues because the samples to be analysed are generally small and the amount of residue is insufficient to allow effective application of infrared, mass, and other spectroscopic methods.

The derivatization technique has been used by a number of workers (Gunther 1962, Klein *et al.* 1964, Hammence *et al.* 1965, Sans 1967, Duffy and Wong 1967, Osadchuk and Wanless 1968, Cochrane and Chau 1968, Chau 1969, Chau and Cochrane 1969, Wiencke and Burke 1969) and is gaining in popularity. Success of the technique usually depends on the specific derivative remaining lipophilic and therefore extractable with the organic solvent (usually hexane) for direct analysis by the usual GLC-EC method.

A number of chemical reactions is available, but two of the more broad-spectrum types that are quite useful for the pesticides generally found in wildlife are:

(i) Reaction with ethanolic potassium hydroxide (mainly dehydrochlorination) to give products with shorter retention times. This is effective for DDT, DDD, and their isomers, as well as α- and γ-BHC; the β-isomer is not affected.

(ii) Reaction with hydrogen bromide/acetic anhydride reagent to give bromohydrin or bromoacetoxy derivatives with longer retention times. This reaction has been found useful for heptachlor, aldrin, heptachlor epoxide, dieldrin, and endrin.

c) Thin-layer chromatography. In spite of certain drawbacks with TLC for confirmation of small amounts of pesticide residues (Reynolds 1969 a) the technique is still valuable for the confirmation of DDE. TLC can be quite useful also for some group separation, as an ancillary cleanup technique, and for semiquantitative estimation of some pesticide residues.

IX. Identification of PCB's in samples

Data obtained by our laboratory in the course of work conducted for The Canadian Wildlife Service have demonstrated that PCB's are present in some but not all species of Canadian wildlife. Undoubtedly the presence of PCB's has complicated the pesticide residue methodology, but this complication cannot be avoided if meaningful results are to be obtained. Since not all samples contain PCBs, it would be unwise to do a Florisil separation on all extracts before having some indication whether or not the separation is necessary. Accordingly our laboratory uses the approach described below.

The sample to be analysed is extracted, cleaned-up, and assayed using the SE-30/QF-1 column. Depending on the EC results, a decision is made whether or not to make a Florisil separation and check further for the presence of PCB's. This decision is based on the assumption that if there are no apparent *p,p'*-DDD and *p,p'*-DDT peaks, then the presence of PCB's would not be expected. The DDD-DDT combination is chosen as the criterion because peaks eight and ten, two of the larger peaks in Aroclor 1254 and 1260 have identical retention times with *p,p'*-DDD and *p,p'*-DDT, respectively. Also, the PCB components giving rise to peaks eight and ten are less likely to be metabolized than the earlier emerging compounds. Hence, if PCB's (1254 or 1260 pattern) are present, peaks should show up giving apparent *p,p'*-DDD and *p,p'*-DDT values with the EC detection system.

Other useful information regarding the presence of PCB's can be obtained from the initial gas chromatography. If the GLC pattern (prior to PCB separation on Florisil) shows high "apparent" *p,p'*-DDE with little or no *p,p'*-DDD and *p,p'*-DDT present, then all or most of the "apparent" DDE is probably "true" DDE. This is because peak five (the peak which interferes with *p,p'*-DDE) in the common PCB commercial mixtures (Aroclor 1254 and 1260) is relatively small compared to the DDD- and DDT-interfering peaks (eight and ten) as shown in Figure 1. This rule has been borne out by TLC confirmation of DDE in a number of such cases.

As mentioned above, the two most commonly found PCB types in wildlife samples resemble Aroclor 1254 and 1260. Although these two Aroclors are quite similar in many respects, certain differences in their GLC patterns, as shown in Figure 3, can be used to identify one from the other. With a higher chlorine content, Aroclor 1260 shows about 17 major peaks (SE-30/QF-1 column) compared to 14 with Aroclor 1254. Besides this, one other obvious major difference is the peak-height ratio of peaks ten and 13. This ratio has rough values of 9.3 and 1.0 for Aroclors 1254 and 1260, respectively.

Additional information regarding the PCB mixture present in a sample can be obtained by comparisons of other peak-height ratios of the PCB standards and the PCB's in the sample. However, it must be borne in mind that in practice one might be faced with complex mixtures of PCB's rather than a single, specific mixture. In such cases an average value seems to be the best compromise. The method of estimating PCB's described below attempts to take this into account.

Although most of the early reports on the presence of PCB's in samples indicated the Aroclor 1254 GLC pattern, more recent work by KOEMAN *et al.* (1969) and the author's laboratory has indicated that a proportion of the specimens does contain the Aroclor 1260 GLC pattern. This apparent predominance of the Aroclor 1254 type might have been due to its availability as a standard and lack of the Aroclor 1260 type in the early developmental stages of PCB methodology. The GLC profiles of a number of commercially available PCB mixtures were determined using the SE-30/QF-1 column. These are shown in Figure 4 and indicate that the more highly chlorinated

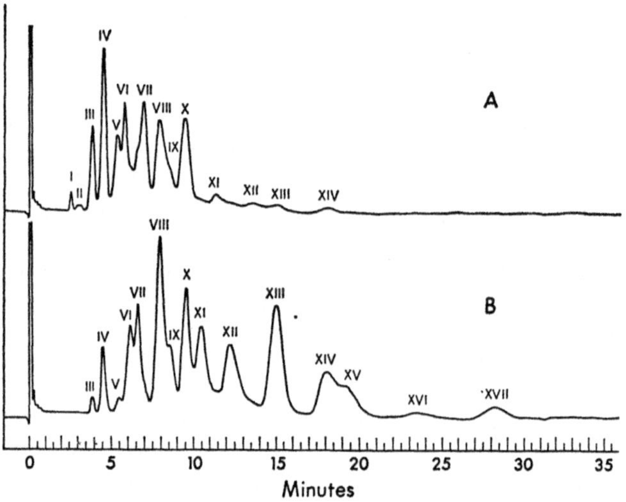

Fig. 3. Comparison of Aroclors 1254 and 1260. Chromatogram *A:* 5 ng. of Aroclor 1254; the 14 major peaks are numbered I to XIV as in chromatogram *B* of Figure 1. Chromatogram *B:* 5 ng. of Aroclor 1260; the 17 major peaks are numbered I to XVII (the early peaks correspond to those in Aroclor 1254). GLC parameters as in Figure 1

biphenyls (Aroclors 1254 and 1260) are easily detected with the usual operating parameters. On the other hand, the compounds of the lower chlorinated mixtures (Aroclors 1221, 1232, and 1242) and the higher molecular weight mixtures (e.g., Aroclor 5460) are less responsive and are, therefore, more likely to go undetected under normal operating conditions. Apparently, the lowered sensitivity is due to low chlorine content in the first group and to high molecular weight with resultant long retention times in the second group. In both cases, therefore, higher concentration would be a prerequisite to ensure detection under normal conditions. This, as well as the possibility of metabolism of the less highly chlorinated biphenyls (KOEMAN *et al.* 1969), might explain the general absence or rarity of these compounds in wildlife samples.

It must not be overlooked, therefore, that in the presence of large amounts of the lower chlorinated biphenyls (e.g., Aroclors 1221, 1232,

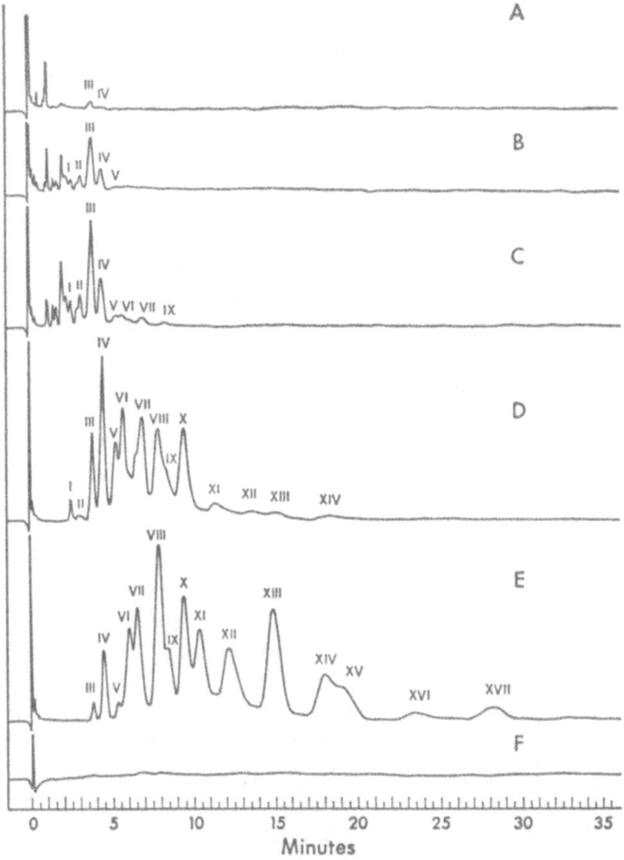

Fig. 4. GLC profiles of the more popular Aroclor mixtures under normal analytical conditions. Note: All peak numbers correspond to those of Aroclor 1254. Chomatograms: *A* = 5 ng. of Aroclor 1221, *B* = 5 ng. of Aroclor 1232, *C* = 5 ng. of Aroclor 1242, *D* = 5 ng. of Aroclor 1254, *E* = 5 ng. of Aroclor 1260, and *F* = 5 ng. of Aroclor 5460. GLC parameters as in Figure 1

and 1242) and negligible quantities of the higher chlorinated biphenyls (e.g., Aroclors 1254 and 1260), different criteria would be necessary to detect PCB interference. To date, this condition has not been encountered in Canadian wildlife samples.

As might be expected from the complex nature of the PCB mixtures and the chemical inertness of the individual compounds, the confirmation of identities of these materials is much more difficult than with the organochlorine pesticides. Thus, in the absence of a mass spectrometer, confirmation of the PCB's is mainly by their GLC retention times and patterns on the mixed and polar phase columns in conjunction with their non-reactivities with the two general reagents used for the organochlorine pesticides.

The prior separation of the PCB's from the organochlorine pesticides by the Florisil technique also aids in the identification of the PCB's.

X. Estimation of PCB's

Since the PCB's are themselves toxic, attempts have been made to estimate the amounts present in samples. It should, however, be recognized that only rough estimates can be made since pure standards of the individual compounds in the mixtures are to date unavailable. The situation resembles that of toxaphene estimation, but is of greater complexity.

KOEMAN *et al.* (1969) based their PCB estimation on the peak of phenoclor DP6 having r_x (relative retention time with dieldrin = 1 under his operating conditions) equal to 1.45. RISEBROUGH *et al.* (1969) estimated the PCB's on the assumption that they have similar EC responses to *p,p'*-DDE and applied a factor to fit the assumed 54 percent chlorine content of the PCB's. JENSEN *et al.* (1969) reported estimates as the sum of all the PCB components.

The author's laboratory uses a method very similar to KOEMAN's with the exception that an average based on two peaks is usually employed. Should the two values before averaging vary to any extent, additional estimations are made using other peaks, and all the results are averaged. In the case of a few samples, the PCB level was obtained by averaging results from all the major peaks in the mixture; however, the use of peaks eight and ten only has been found to give equally satisfactory results. Note that results are reported as being based on Aroclor 1254 or 1260 depending on the overall GLC pattern. Peaks eight and ten were chosen because of their stability (see section on identification). In a later section the above points are illustrated by analysis of resin powder taken from a fish hatchery trough.

The total PCB results obtained by the method described may be slightly higher than the actual values because the early emerging PCB peaks (mainly one, two, and three) are usually absent from the chromatograms of sample extracts. However, this should not affect the overall results greatly since these peaks represent minor components in the mixtures. Furthermore, there might be compensation, since no attempt has been made to apply recovery correction factors.

The method of estimation described above appears to give a reasonable estimate of PCB contamination based on commercial PCB mixtures. For absolute measurement of PCB content it would be necessary to prepare standards for the individual PCB components, and, even if these were available, the determinations would be complex and costly. Considering our present limited knowledge of total PCB toxicity and our even greater ignorance about the toxicity of the individual components of the mixtures, it is doubtful whether such accurate determinations would be meaningful at this time. As stressed earlier, the most urgent need is for toxicological investigation in this area to indicate whether some of the PCB compounds are more toxic or have greater sublethal effect than others, as observed with the BHC isomers. At that point it would become essential to identify and estimate accurately specific individual compounds in the mixtures.

XI. Separation of PCB's from organophosphorus compounds by the Florisil technique

A limited amount of work has been carried out in our laboratory to determine whether the Florisil column technique is applicable to the separation of the PCB's from the organophosphorus compounds. Preliminary results on those tested (phorate, diazinon, ronnel, malathion, parathion, methyl Trithion, ethion) indicate that they are not eluted with the PCB's by hexane, and can therefore be separated. In comparison to the organochlorine pesticides, the organophosphorus compounds generally require larger volumes of the ether-hexane mixture for elution from the Florisil column, most likely because of their more polar nature.

As with the organochlorine pesticides, the presence of PCB's in a sample extract will interfere with some of the more common organophosphates when the EC detection system is used. However, with the use of a dual detector (EC + phosphate or thermionic) connected to a dual-pen recorder, the effluent from a single injection containing PCB's and organophosphates can be split and differentiation made as shown in Figure 5. Like the organochlorine pesticides, the PCB's are not detected on the phosphate detector. Nevertheless, in the absence of a detector which is specific for phosphorus, the Florisil column technique can be used to separate the PCB's from the organophosphates to facilitate the analyses.

XII. Residues of organochlorine pesticides and PCB's in Canadian wildlife

Following the approaches outlined, the author's laboratory has analysed a number of Canadian wildlife specimens for residues of

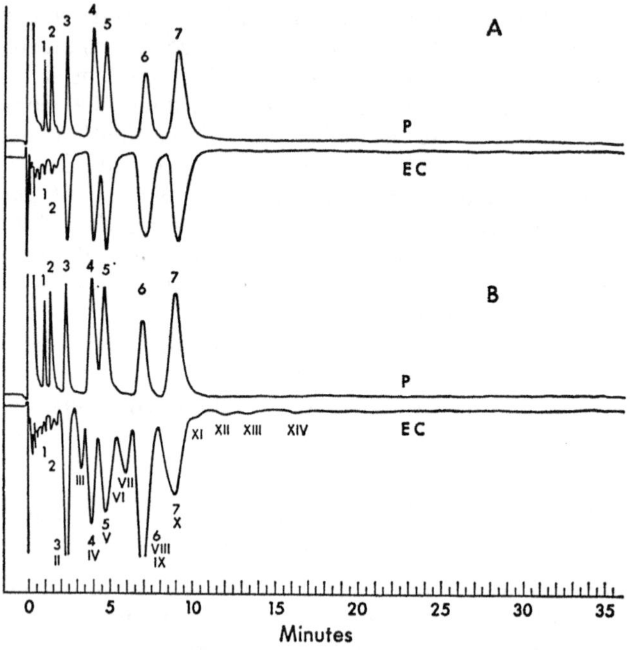

Fig. 5. Differentiation of Aroclor 1254 and organophosphorus pesticides using dual detectors and effluent split. Chromatogram *A:* organophosphorus standard pesticide mixture; peak numbers 1 = 1.75 ng. of phorate, 2 = 2.0 ng. of diazinon, 3 = 3.0 ng. of Ronnel, 4 = 13.0 ng. of malathion, 5 = 5.0 ng. of parathion, 6 = 15.0 ng. of methyl Trithion, and 7 = 10.0 ng. of ethion. Chromatogram *B:* combination of organophosphorus mixture (chromatogram *A*) and 10 ng. of Aroclor 1254 numbered I to XIV as in chromatogram *B* of Figure 1 (a compromise of the E.C. response is necessary to obtain maximum response on the phosphorus detector). *P* = phosphorus detector and *EC* = electron capture detector. Column four percent OV-100/six percent OV-210 on 60/80 mesh Chromosorb W (AW); other GLC parameters as in Figure 1

organochlorine pesticides and PCB's. Some of the data on aquatic birds from Western Canada are given in Table I. The inclusion of these tabulated results, which represent only a minor portion of the total number of samples analysed, is intended to illustrate the types of analytical problems created by the presence of PCB's, and to indicate potential errors due to high levels of PCB's in samples.

Referring to Table I, some general remarks are in order:

1. The greater portion of the apparent DDE in samples is in fact DDE. This is predictable from the comparatively smaller amounts of apparent DDD and DDT, and has been verified by TLC.
2. A large proportion of the apparent DDD in the samples is due to PCB's. However, DDD is present in some samples.

3. The effect of PCB contamination on DDT values is less marked than in the case of DDD, but it is still substantial.
4. The apparent HE values are in most cases due entirely to the pesticide. This is somewhat incongruous with the rest of the data since peak four of the commercial PCB mixtures does interfere with HE. However, this anomaly may be explained on the basis that the PCB component giving rise to peak four is degraded extensively in the bird's metabolic processes.
5. Based on all of the PCB estimates in wildlife samples noted to date, including those in Table I, the following general trends are evident:
   (a) Aquatic raptorials generally contain higher residue levels of pesticides and PCB's than other wildlife species.
   (b) High levels of PCB's are found only in the presence of high levels of organochlorine pesticides (the resin powder mentioned below is an obvious exception) but high levels of organochlorine pesticides are not necessarily accompanied by high levels of PCB's.
   (c) The most abundant PCB types detected are those which show Aroclor 1254 and 1260 patterns and, therefore, fit well with the proposed method of quantitation. However, it should not be overlooked that with the present residue methodology some other PCB-type compounds might go undetected.
6. It is important to measure moisture and fat content of samples in conjunction with residue levels. This allows for reporting on a wet-, dry-, or fat-weight basis. However, caution should be used when reporting residues in wildlife samples, for example, on a fat basis when the fat content is very low. Such figures can be inaccurate as well as highly misleading.

The gas chromatogram of a typical sample showing the presence of PCB's is given in Figure 6. It indicates the Aroclor 1260 pattern and was obtained from a duckling found dead in the Toronto area in the summer of 1969.

A good example of PCB's showing up in unexpected places occurred recently. A wildlife biologist was faced with the problem of fish dying in a hatchery for no apparent reason. He scraped some of the resin powder from the hatchery trough and found, surprisingly, that experimental fish were susceptible to even a dilute solution of the powder. He submitted a sample of the resin powder to the author's laboratory, and upon analysis a typical GLC pattern of PCB's (Aroclor 1254 type) was obtained. This is shown in Figure 7. Using the methodology described earlier in this review, the sample was found to contain PCB's but no pesticides. Based on calculations from 11 individual peaks of Aroclor 1254, the PCB concentration was estimated at 35 p.p.m (range 25 to 46). Using the average results based on peaks eight and ten only, the PCB estimate was 29 p.p.m. This is

Table I. *Organochlorine residues (p.p.m. on a wet-weight basis) in Canadian aquatic birds as measured before and after PCB Florisil separation*

Description of specimen[a]	p,p'-DDE[b]	Dieldrin[c]	p,p'-DDD			p,p'-DDT			HE			PCB[e]
			Before (apparent)	After	%[d]	Before (apparent)	After	%[d]	Before (apparent)	After	%[d]	
California Gulls												
Abd. fat bird 1	241	1.36	3.16	ND[f]	0	3.17	ND	0	ND	ND	—	50.8
Abd. fat bird 2	138	1.02	1.02	ND	0	1.04	0.14	7	0.23	0.22	96	25.3
Abd. fat bird 3	240	1.23	2.63	ND	0	3.47	0.25	7	0.19	0.21	107	62.6
Abd. fat bird 4	246	1.91	1.59	ND	0	2.41	1.27	53	ND	ND	—	36.7
Abd. fat bird 5	169	1.78	1.71	ND	0	2.05	0.35	17	0.33	0.19	58	23.2
Abd. fat bird 6	296	1.51	3.09	ND	0	3.33	1.09	33	0.42	0.36	87	23.0
Brain bird 7	3.6	0.04	0.04	ND	0	0.04	ND	0	ND	ND	—	0.81
Liver bird 7	15.7	0.07	0.11	0.02	14	0.11	0.01	2	0.03	0.03	79	1.69
Ovary bird 7	22.9	0.12	0.17	0.04	24	0.23	0.02	8	0.05	0.04	83	2.52
Abd. fat bird 7	416	1.93	4.64	0.34	7	4.43	0.62	14	1.19	0.78	65	38.4
Egg area 1	21.8	0.19	0.20	ND	0	0.25	0.08	33	0.07	0.05	66	1.74
Egg area 2	2.9	0.09	0.11	ND	0	0.11	0.02	18	0.01	0.01	93	1.57
D.C. Cormorant Eggs												
Sample 1 area 3	3.3	0.11	0.11	ND	0	0.12	0.03	22	0.01	0.01	100	1.94
Sample 2 area 3	3.3	0.10	0.12	ND	0	0.12	0.02	13	0.03	0.03	110	2.21
10 pooled area 4	3.5	0.21	0.03	ND	0	0.07	0.03	41	0.03	0.04	106	0.67
Mallard Duck Eggs												
10 pooled area 4	0.2	0.08	0.01	0.01	80	0.15	0.15	100	0.01	0.01	100	0.09
Common Tern Eggs												
Sample 1 area 5	33.3	0.15	0.17	0.02	13	0.26	0.15	51	ND	ND	—	1.57
Sample 2 area 5	5.1	0.02	0.06	ND	0	0.16	0.15	90	ND	ND	—	0.94
Great Blue Heron Eggs												
Sample 1 area 5	78.0	0.05	0.27	0.27	100	0.41	0.42	102	0.02	0.02	94	Trace[g]

Table I. (continued)

Description of specimen[a]	p,p'-DDE[b]	Dieldrin[c]	p,p'-DDD			p,p'-DDT			HE			PCB[e]
			Before (apparent)	After	%[d]	Before (apparent)	After	%[d]	Before (apparent)	After	%[d]	
Sample 1 area 6	24.0	0.42	0.33	0.27	81	0.18	0.10	57	0.15	0.12	83	3.41
Sample 2 area 6	11.3	0.19	0.07	ND	0	0.30	0.23	77	0.10	Trace[g]	—	1.93
Sample 3 area 6	13.7	0.74	1.63	1.70	104	0.08	0.06	69	0.05	0.05	100	1.09
Sample 1 area 7	13.5	0.07	0.07	0.05	64	0.05	0.02	33	0.02	Trace[g]	—	0.57
Sample 2 area 7	3.7	0.06	0.12	ND	0	0.14	0.03	18	0.02	Trace[g]	—	1.38
Sample 1 area 8	6.3	0.68	0.09	0.05	49	0.04	0.01	22	0.02	0.01	52	0.53
Sample 2 area 9	9.1	0.07	0.06	0.03	52	0.07	0.05	70	0.03	0.01	44	0.33

[a] These specimens are from western Canada (Vermeer and Reynolds 1970).

[b] These are "Before PCB-Florisil separation" values and are likely to include PCB contributions. The "After" values (which do not change significantly from the "Before" values) are not given because the PCB peak five (the DDE-interfering peak) of Aroclor 1254 or 1260 is not separated from DDE by this method, but, as discussed earlier, the PCB contributions to the DDE values are expected to be relatively small compared to the contributions to DDD or DDT.

[c] These are "Before" values. The "After" values (not given) are not significantly different from the "Before" values since there is no direct PCB interference. However, in the presence of high levels of DDE, low levels of dieldrin will be masked. In such cases a separation (e.g., PCB-Florisil technique) is necessary, and the "After" value would be reported as the dieldrin level.

[d] $\% = \frac{\text{After value}}{\text{Before value}} \times 100$ and indicates the percentage of the total apparent pesticide level that was actually due to the pesticide.

[e] Values (p.p.m.) based on peaks eight and ten of Aroclor 1254.

[f] ND = none detected = <0.0001 p.p.m.

[g] Trace = <0.001 p.p.m.

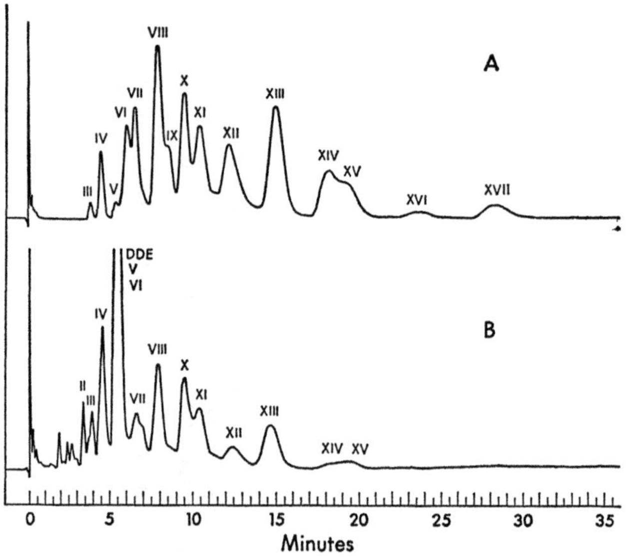

Fig. 6. Typical sample indicating the presence of PCB's of the Aroclor 1260 type in a duckling from the Toronto area. Chromatogram *A:* 5 ng. of Aroclor 1260 numbered I to XVII as in chromatogram *B* of Figure 3. Chromatogram *B:* the equivalent of 0.35 mg. of duck sample (hexane portion of Florisil split). Note the DDE contribution to PCB peak V. GLC parameters as in Figure 1

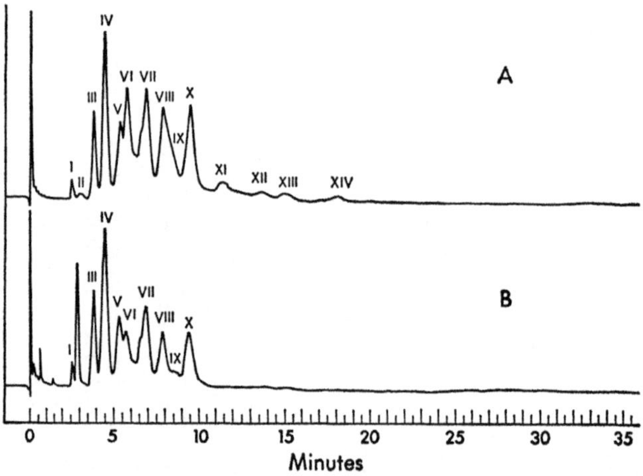

Fig. 7. Resin powder extract indicating the presence of PCB's of the Aroclor 1254 type. Chromatogram *A:* 5 ng. of Aroclor 1254 numbered I to XIV as in chromatogram *B* of Figure 1. Chromatogram *B:* The equivalent of 0.02 mg. of resin powder extract (hexane portion of Florisil split), numbered I to XIV as in Aroclor 1254. GLC parameters as in Figure 1

an interesting example of PCB's causing contamination as a direct result of industrial application.

With regard to the Florisil technique for the separation of PCB's from pesticides, one important modification to the original method (REYNOLDS 1969) is now used routinely in the wildlife studies. Originally the scheme was designed to accommodate simultaneously both the cleanup of the sample extract and the separation of PCB's. However, as pointed out earlier in this review, not all samples contain PCB's. Consequently, for general samples, to avoid unnecessary splitting of all sample extracts (each split doubles the number of eluates to be chromatographed), a normal cleanup including a Florisil column, but no differential elution, is conducted. The Florisil-PCB separation is then carried out on the cleaned-up extract, if it is required. Under these conditions smaller columns have been found to be as efficient as the larger columns used in the original method, and there is a substantial saving in the amount of reagents required. Details of the modified procedure are as follows:

A glass column (44 cm. × 1 cm. i.d. having a 50-ml. reservoir on top) is packed with 30 cm. (~10.7 g.) of Florisil (60/100 mesh, Floridin Co., pesticide grade, stored at 130°C. until ready for use) and topped with a 2-cm. layer of anhydrous Na_2SO_4. The sample extract of about 5 ml. (*n*-hexane) is added to the column. The first elution which removes PCB's, heptachlor, aldrin, and DDE, is effected with 60 ml. of *n*-hexane (Nanograde, Mallinckrodt). The remaining pesticides are then eluted with 40 ml. of 50 percent ethyl ether in hexane.

Note that in filling the columns, the Florisil is packed down by gentle tapping, and the sample extract is added to a dry column, *i.e.*, no prewetting. After the first elution the receiver is changed as soon as the last of the hexane meets the Na_2SO_4. It should be noted also that with one batch of pesticide grade Florisil, some PCB peaks (two, six, nine, 11) showed partial elution in the second eluate. These, however, are comparatively minor peaks and do not interfere with any of the common organochlorine pesticides. As there are differences in batches of Florisil, the procedure has to be checked out fully prior to application to actual samples. Modifications of volumes, etc., may be necessary to effect the desired separation. It is likely that adsorbents other than Florisil, e.g., silica gel or alumina, will give similar separation of PCB's from pesticides but these have not been investigated to date.

Acknowledgments

Work in connection with this review was supported by funds from the Pesticide Section, Canadian Wildlife Service, Ottawa, and from the Province of Ontario through the Department of Trade and Development. The writer gratefully acknowledges helpful suggestions of Dr. S. G. Reid, Director of Organic Chemistry Department, Ontario

Research Foundation in preparing this review for publication. The technical assistance of Terry Cooper and Sandra Takacs is gratefully acknowledged.

Table II. *Chemical designations of pesticides mentioned in text*

Pesticide	Chemical name
aldrin	1,2,3,4,10,10-hexachloro-1,4,4a,5,8,8a-hexahydro-1,4-*endo*, *exo*-5,8-dimethanonaphthalene
p,p'-DDD (TDE)	2,2-bis(*p*-chlorophenyl)-1,1-dichloroethane
p,p'-DDE	1,1-dichloro-2,2-bis(*p*-chlorophenyl)ethylene
p,p'-DDT	1,1,1-trichloro-2,2-bis(*p*-chlorophenyl)ethane
diazinon	*O,O*-diethyl-*O*-(2-isopropyl-4-methyl-6-pyrimidyl)-phosphorothioate
dieldrin	1,2,3,4,10,10-hexachloro-6,7-epoxy-1,4,4a-5,6,7,8,8a-octahydro-1,4-*endo,exo*-5,8-dimethanonaphthalene
ethion	*O,O,O',O'*-tetraethyl-*S,S'*-methylene bisphosphorodithioate
heptachlor	1,4,5,6,7,8,8-heptachloro-3a,4,7,7a-tetrahydro-4,7-*endo*-methanoindene
heptachlor epoxide	1,4,5,6,7,8,8-heptachloro-2,3-epoxy-3a,4,7,7a-tetrahydro-4, 7-methanoindane
lindane (γ-BHC)	γ-1,2,3,4,5,6-hexachlorocyclohexane
malathion	*S*-[1,2-bis(ethoxy carbonyl)ethyl]*O,O*-dimethyl phosphorodithioate
methyl Trithion	*O,O*-dimethyl *S*-(*p*-chlorophenylthio) methyl phosphorodithioate
parathion	*O,O*-diethyl *O*-*p*-nitrophenyl phosphorothioate
phorate (Thimet)	*O,O*-diethyl *S*-(ethylthio) methyl phosphorodithioate
Ronnel	dimethyl 2,4,5-trichlorophenyl phosphorothionate

Summary

Since 1966, PCB's have been detected in fish and wildlife, especially in Europe and North America. Contamination from industrial wastes via the water ways appears to be the main source of PCB's in aquatic species. The PCB's have many important industrial uses but are not utilized as pesticides.

Due to their similarities in structure and properties to the organochlorine pesticides, the PCB's tend to interfere with accurate GLC-EC determination of organochlorine and many organophosphorus pesticides. In addition to their interference with pesticide residue analysis the PCB's are themselves toxic so that their quantitation is desirable.

A method for recognising the presence of PCB's in samples, separating them from pesticides, and determining the pesticide levels accurately is described. Simultaneously an estimate of total PCB content is afforded.

The residue data of some Canadian aquatic birds indicate that high levels of PCB's are found only in the presence of correspondingly high levels of the DDT pesticide group. On the other hand, high organochlorine pesticide levels are not necessarily accompanied by high PCB content.

Further research on toxicity and sublethal effects of PCB mixtures and their individual components is required. If particular components are shown to be more toxic than others, refining of the PCB quantitation methods will be necessary

Résumé*

Analyse de résidus de pesticides en présence de diphényls polychlorés

Depuis 1966 des diphényls polychlorés (PCBs) ont été détectés dans les poissons et le gibier, particulièrement en Europe et en Amérique du Nord. Il apparait que la présence de PCBs dans la faune aquatique soit due principalement à la contamination des cours d'eau par des déchets industriels. Les PCBs sont utilisés pour un grand nombre de procédés industriels importants, mais ils ne sont pas appliqués comme pesticides.

Etant donné que la structure et les propriétés des PCBs sont semblables à celles des pesticides organochlorés, ils ont tendance à fausser la détermination exacte des substances organochlorées et organophosphorées lors d'analyses par chromatographie en phase gazeuse via capture d'électrons. En plus de leur interférence lors de l'analyse des résidus des pesticides, les PCBs sont eux-mêmes des composés toxiques dont la détermination quantitative est souhaitable.

Le présent article contient une méthode qui permet de reconnaître la présence des PCBs dans des échantillons et, en même temps, de les séparer des pesticides dont le taux de résidus peut être exactement déterminé. La méthode permet également une estimation du contenu total en PCBs.

Des résultats d'analyses portant sur des oiseaux aquatiques canadiens indiquent qu'un taux élevé de PCBs est lié à un taux élevé de résidus de pesticides du groupe DDT. D'autre part, des taux élevés de résidus de substances organochlorées ne sont pas nécéssairement accompagnés d'un taux élevé de PCBs.

Des recherches supplémentaires sur la toxicité et les effets sublétaux des mélanges de PCBs et de leurs composés individuels seront nécéssaires. Si on peut démontrer que des substances particulières sont plus toxiques que d'autres, on aura besoin d'une méthode plus spécifique permettant l'analyse individuelle des PCBs.

---

* Traduit par H. Geissbühler.

Zusammenfassung*

Die Analyse von Pflanzenschutzmittel-Rückständen in Gegenwart von Polychlordiphenylen

Seit 1966 werden Polychlordiphenyle (PCB-Stoffe) in Fischen und Wildtieren, besonders in Europa und Nordamerika, nachgewiesen. Die Verschmutzung der Wasserwege durch Industrieabfälle scheint dabei die Hauptursache von PCB-Rückständen in der aquatischen Fauna zu sein. Die PCB-Stoffe werden für eine Anzahl wichtiger Industrieprozesse gebraucht, als Schädlingsbekämpfungsmittel werden sie jedoch nicht eingesetzt.

Da die PCB-Stoffe in ihrer Struktur und in ihren Eigenschaften den chlorierten Kohlenwasserstoff-Insektiziden sehr ähnlich sind, können sie die gaschromatographische Bestimmung mit Elektronen-einfang-Detektor von chlorierten Kohlenwasserstoffen und mancher Phosphorsäureester verfälschen. Die PCB-Stoffe können nicht nur Rückstandsanalysen beeinträchtigen, sondern sie sind selbst toxische Substanzen, deren Nachweis sich aufdrängt.

Im vorliegenden Artikel wird eine Methode beschrieben, die es erlaubt, das Vorhandensein von PCB-Stoffen in Proben nachzuweisen, sie von den Schädlingsbekämpfungsmitteln abzutrennen und die Rückstände der letztern genau zu bestimmen. Darüber hinaus gestattet die Methode, den Totalgehalt an PCB-Stoffen abzuschätzen.

Rückstandsdaten einiger kanadischer Wasservögel weisen darauf hin, dass hohe PCB-Konzentrationen nur in Gegenwart hoher Rückstände von Substanzen der DDT-Gruppe auftreten. Anderseits sind hohe Rückstände von chlorierten Kohlenwasserstoff-Insektiziden nicht unbedingt von hohen PCB-Konzentrationen begleitet.

Weitere Untersuchungen über die Toxizität und die sublethale Wirkung von PCB-Mischungen und ihrer individuellen Komponenten sind angezeigt. Sollte es sich dabei ergeben, dass gewisse Einzelstoffe toxischer sind als andere, wird eine Verfeinerung der Nachweis- und Bestimmungsmethoden für PCB-Stoffe notwendig.

References

BLOCK, S. S.: Insecticidal surface coatings. Soap & Sanit. Chem. **24** (2),138 (1948).

BROWN, R. M.: On the toxicity of the "Aroclors". Chemist Analyst **36**, 33 (1947).

BURDICK, G. E., E. J. HARRIS, H. J. DEAN, T. M. WALKER, J. SKEA, and D. COLBY: The accumulation of DDT in lake trout and the effect on reproduction. Trans. Amer. Fisheries Soc. **93** (2), 127 (1964).

BURKE, J. A.: Gas chromatography for pesticide residue analysis; some practical aspects. J. Assoc. Official Anal. Chemists **48**, 1037 (1965).

CHAU, A. S. Y.: Derivative formation for the confirmation of endosulfan by gas chromatography. J. Assoc. Official Anal. Chemists **52**, 1240 (1969).

* Übersetzt von H. GEISSBÜHLER.

———, and W. P. COCHRANE: Cyclodiene chemistry. III. Derivative formation for the identification of heptachlor epoxide, cis-chlordane, trans-chlordane, dieldrin, and endrin pesticide residues by gas chromatography. J. Assoc. Official Anal. Chemists **52**, 1220 (1969).

COCHRANE, W. P., and A. S. Y. CHAU: Note on gas chromatographic identification of heptachlor pesticide residues by derivative formation. J. Assoc. Official Anal. Chemists **51**, 1267 (1968).

DUBOIS, K. P.: Combined effects of pesticides. In: Current views on pesticides, p. 35. Symposium sponsored by Food and Drug Directorate, *Department of National Health and Welfare*, Ottawa, 1968; reprinted from Can. Med. Assoc. J. **100** (No. 4) (Jan. 25, 1969).

DUFFY, J. R., and N. WONG: Residues of organochlorine insecticides and their metabolites in soils in the Atlantic provinces of Canada. J. Agr. Food Chem. **15**, 457 (1967).

GREENBURG, L., M. R. MAYERS, A. R. SMITH: The systemic effects resulting from exposure to certain chlorinated hydrocarbons. J. Ind. Hyg. Toxicol. **21**, 29 (1939).

GUNTHER, F. A.: Instrumentation in pesticide residue determinations. Adv. Pest Control Research **5**, 191 (1962).

HAMMENCE, J. H., P. S. HALE, and D. J. CAVERLY: The identification and determination of chlorinated pesticides residues. Analyst **90**, 649 (1965).

HARRISON, R. B.: Residues in wildlife with special reference to endrin. J. Sci. Food Agr. **17**, 10 (1966).

HICKEY, J. J., and D. W. ANDERSON: Chlorinated hydrocarbons and eggshell changes in raptorial and fish-eating birds. Science **162**, 271 (1968).

HOLDEN, A. V., and K. MARSDEN: The examination of surface waters and sewage effluents for organochlorine pesticides. J. Proc. Inst. Sewage Purif. p. 295 (1966).

——— ——— Organochlorine pesticides in seals and porpoises. Nature **216**, 1274 (1967).

HOLMES, D. C., J. H. SIMMONS, and J. O'G. TATTON: Chlorinated hydrocarbons in British wildlife. Nature **216**, 227 (1967).

HORNSTEIN, I., and W. N. SULLIVAN: The role of chlorinated polyphenyls in improving lindane residues. J. Econ. Entomol. **46**, 937 (1953).

JENSEN, S.: Report of a new chemical hazard. New Scientist **32**, 612 (1966).

———, A. G. JOHNELS, M. OLSSON, and G. OTTERLIND: DDT and PCB in marine animals from Swedish waters. Nature **224**, 247 (1969).

———, and G. WIDMARK: Swedish report at the OECD pesticide conference on "Unintended residues in the environment" (1967).

JONES, J. W., and H. S. ALDEN: An acneform dermatergosis. Arch. Dermatol. Syphilol. **33**, 1022 (1936).

KLEIN, A. K., and J. O. WATTS: Separation and measurement of perthane, DDD (TDE), and DDT in leafy vegetables by electron capture gas chromatography. J. Assoc. Official Anal. Chemists **47**, 311 (1964).

KOEMAN, J. H., A. A. G. OSKAMP, J. VEEN, E. BROUWER, J. ROOTH, P. ZWART, E. V. D. BROEK, and H. VAN GENDEREN: *Meded. Rijks-faculteit Landbouwwetenschappen* Gent. **32**, 841 (1967).

KOEMAN, J. H., M. C. TEN NOEVER DE BRAU, and R. H. DE VOS: Chlorinated biphenyls in fish, mussels and birds from the River Rhine and the Netherlands coastal area. Nature **221**, 1126 (1969).

LINDQUIST, A. W., A. H. MADDEN, H. G. WILSON, and E. F. KNIPLING: DDT as a residual-type treatment for control of house flies. J. Econ. Entomol. **38**, 257 (1945).

MCLAUGHLIN, J., JR., J. P. MARLIAC, M. J. VERRETT, M. K. MUTCHLER, and O. G. FITZHUGH: The injection of chemicals into the yolk sac of fertile eggs prior to incubation as a toxicity test. Toxicol. Applied Pharmacol. **5**, 760 (1963).

Miller, J. W.: Pathologic changes in animals exposed to a commercial chlorinated diphenyl. U. S. Public Health Reports **59**, 1085 (1944).

Monsanto Co.: The Aroclor compounds, p. 17 (1965).

——— Aroclor plasticizers. Tech. Bull. O/PL-306 (1967).

Osadchuck, M., and E. B. Wanless: Identification and quantitative estimation of aldrin residues in the presence of interfering materials in samples of plant and animal origin. J. Assoc. Official Anal. Chemists **51**, 1264 (1968).

Peakall, D. B.: Pesticide-induced enzyme breakdown of steroids in birds. Nature **216**, 505 (1967).

Penning, C. H.: Physical characteristics and commercial possibilities of chlorinated biphenyl. Ind. Eng. Chem. **22**, 1180 (1930).

Ratcliffe, D. A.: Decrease in eggshell weight in certain birds of prey. Nature **215**, 208 (1967).

Reichel, W.: Personal communication (1969).

Reynolds, L. M.: Canadian wildlife service reports (Unpublished data) (1968, 1969).

——— Pesticide residue analysis in the presence of PCB's. 4th Annual W. Can. Seminar on Pesticide Residue Analysis. Winnipeg, June 2–4 (1969 a).

——— Polychlorobiphenyls (PCB's) and their interference with pesticide residue analysis. Bull. Environ. Contamination Toxicol. **4**, 128 (1969 b).

Risebrough, R. W., P. Rieche, S. G. Herman, D. B. Peakall, and M. N. Kirven: Polychlorinated biphenyls in the global eco-system. Nature **220**, 1098 (1968).

———, P. Rieche, and H. S. Olcott: Current progress in the determination of the polychlorinated biphenyls. Bull. Environ. Contamination Toxicol. **4**, 192 (1969).

Robinson, J.: Residues of organochlorine insecticides in dead birds in the United Kingdom. J. Chem. Ind., p. 1974 (1967).

Roburn, J.: A simple concentration-cell technique for determining small amounts of halide ions and its use in the determination of residues of organochlorine pesticides. Analyst **90**, 467 (1965).

Samuel, B. L., and H. K. Hodges: Insecticide screening methods for organochlorine and organophosphate insecticides in foods and feeds. Residue Reviews **17**, 35 (1967).

Sans, W. W.: Multiple insecticide residue determination using column chromatography, chemical conversion, and gas liquid chromatography. J. Agr. Food Chem. **15**, 192 (1967).

Schmidt, H., and G. Schultz: Patent for manufacture of pentachlorobiphenyl. Ann. **207**, 338 (1881).

Schwartz, L.: An outbreak of halowax acne ("cable rash") among electricians. J. Amer. Med. Assoc. **122**, 158 (1943).

———, and F. A. Barlow: Chloracne from cutting oils. Public Health Rept. **57**, 1747 (1942).

———, and S. M. Peck: Occupational acne. N. Y. State Med. J. **43**, 1711 (1943).

Schechter, M. S., S. B. Soloway, R. A. Hayes, and H. L. Haller: Colorimetric determination of DDT. Ind. Eng. Chem., Anal. ed. **17**, 704 (1945).

Simkiss, K.: Calcium in reproductive physiology. New York: Reinhold (1967).

Stickel, L.: Wisconsin hearings. Science **163**, 551 (1969).

Sullivan, W. N., and I. Hornstein: Chlorinated polyphenyls to improve lindane residues. J. Econ. Entomol. **46**, 158 (1953).

Thompson, J. F., A. C. Walker, and R. F. Moseman: Study of the performance of gas chromatographic columns under severe injection loading. J. Assoc. Official Anal. Chemists **52**, 1251 (1969 a).

——— ——— ——— Evaluation of eight chromatographic columns for chlorinated pesticides. J. Assoc. Official Anal. Chemists **52**, 1263 (1969 b).

TSAO, C. H., W. N. SULLIVAN, and I. HORNSTEIN: A comparison of evaporation rates and toxicity to house flies of lindane and lindane-chlorinated polyphenyl deposits. J. Econ. Entomol. **46**, 882 (1953).

U. S. Food and Drug Administration: Pesticide Analytical Manual, Vol. I and II, revised annually. Washington, D.C. (1969).

VAN TIEL, N.: Improvement of the residual toxicity of DDT solutions by the addition of coumarone resin. Bull. Entomol. Research **43**, 413 (1952).

VERMEER, K., and L. M. REYNOLDS: Organochlorine residues in aquatic birds in the Canadian Prairie Provinces. (In preparation, 1970).

WELCH, R. M., M. LEVINE, and A. H. CONNEY: Effect of chlorinated insecticides on steroid metabolism. In: Chemical fallout. Proc. 1st Rochester Conf. on Environmental Toxicity, Univ. of Rochester (1969).

WIDMARK, G.: IUPAC commission on methods of pesticide residue analysis. J. Assoc. Official Anal. Chemists **50**, 1069 (1967).

WIENCKE, W. W., and J. A. BURKE: Derivatization of dieldrin and endrin for confirmation of residue identity. J. Assoc. Official Anal. Chemists **52**, 1277 (1969).

ZWEIG, G. (ed.): Analytical methods for pesticides, plant growth regulators and food additives, vol. I, p. 1. New York: Academic Press (1963).

Manuscript received 20 January 1970; accepted 17 February 1970.

Development of the Food and Drug Administration's method of analysis for multiple residues of organochlorine pesticides in foods and feeds

By

JERRY A. BURKE*

Contents

I. Foreword**

A method of analysis adequate for the detection and measurement of the appropriate amount of pesticide residues in food products must

* Division of Pesticide Chemistry and Toxicology, Bureau of Foods and Pesticides, Food and Drug Administration, *U. S. Department of Health, Education and Welfare*, Washington, D. C. 20204. The material in this paper was presented to the Commission on Pesticide Residue Analysis, Pesticide Section, International Union of Pure and Applied Chemistry, Cortina, Italy, 1969.

** By J. WILLIAM COOK.*

be available before tolerances are established for those residues in and on food products by the Food and Drug Administration. In most instances the method available at the time the tolerance is established is an individual procedure for one specific chemical. Such methods are useful when it is necessary to analyze food products for those specific chemicals. It became obvious a number of years ago that it was impossible to monitor the levels of pesticides in the general food supply effectively and meaningfully by these individual methods. Since the spray history of a given commodity is generally unknown, the chemist does not know which methods to apply; it is economically and physically impractical to apply a large number of individual methods for all the significant chemical residues that may be present. In addition many pesticide compounds are acted upon by light and air or enzyme systems within plants and animals to produce alteration products which may be significant residues; seldom do individual methods determine these alteration products. On the other hand, the multiresidue methods described later do determine many significant alteration products. It was decided that the only reasonable approach to monitoring the food supply was to simplify analyses by developing methods which measure more than one chemical at a time. An ideal method of analysis would be one which detected and quantitatively measured all of the parent and significant alteration products simultaneously. Such an ideal, of course, is not possible to achieve but a great deal of progress has been made in that direction. It is hoped that even more progress will be forthcoming.

Over a period of 15 to 20 years the chemist has taken advantage of the many new techniques that have become available and has adapted them into the multiresidue analysis system that is discussed later. This multiresidue method is applicable for the tentative identification and quantitative measurement of about 55 organochlorine and about 15 organophosphorus pesticide chemicals and is known to be applicable to about 450 different food products. If any one or a reasonable combination of any of the 70 compounds is present, they can be tentatively identified and quantitatively measured. To aid in the confirmation of the identity of many of the residues there is available a group of confirmatory techniques which are relatively easy to apply. The development of a multiresidue procedure was started before gas-liquid chromatography was commonly available. Since attempts were being made to utilize bioassay and paper chromatography, extraction and cleanup procedures were studied. Many shortcomings of these systems were known at the time gas chromatography became popular. Some of the extraction and cleanup studies that were made in an attempt to improve bioassay and paper chromatographic methods were very helpful in the application of gas chromatography to pesticide residue analysis.

There are advantages to the development and use of these multiresidue procedures aside from the economy of application. The method is sensitive to well below the legal tolerances established for many pesticide chemicals. It is adequately sensitive for the highly toxic compounds so that there is no uncertainty in the determination of many of the very toxic substances. However, since the analysis is made for all compounds simultaneously and the results depend on the chemistry instead of the toxicity, some of the less toxic compounds are measured much below the legal tolerance figure. There are other compounds for which the detector response is poor; therefore these compounds are not measured at a desirable level. These highly sensitive and efficient methods have made it possible to analyze a number of "total diet" or "market basket" samples. High sensitivity is important in these analyses because samples are composites of a number of products, and significant residue amounts in a single component of the sample could be diluted by other components of the composite to very low overall concentrations which may not be measurable. The results of these total diet studies have been helpful to national and international health authorities. Even though these methods are relatively easy to apply to the analysis of food and possibly many other elements in our environment, in order to obtain reliable results it is necessary for the analyses to be conducted (or to be supervised very closely) and raw data to be interpreted by a scientist well trained and experienced in their application and well grounded in the fundamental chemistry of pesticides.

It is not intended to imply that the described methods are the only procedures by which these residues can be detected and measured. The development and validation of this type of scheme is such a tremendous task that it is hoped other scientists will attempt to promote the further development and improvement of this scheme unless, of course, some phenomenal development is achieved which outmodes this type of analysis. It is especially desirable to encourage research on the development of simple, specific, and highly sensitive confirmatory tests which are easily adapted to this scheme.

II. Introduction

Much of the research and development effort of the U. S. Food and Drug Administration (FDA) in pesticide[1] residue analysis has been coordinated toward designing, improving, and extending methods by which a number of residues are detected and measured simultaneously.

These efforts toward development of a comprehensive multiresidue analytical scheme capable of accurate and reproducible determinations have been a continuous process since the beginning of analyses

[1] Pesticides mentioned in text are identified by chemical name in Table IV.

for multiple pesticide residues. A review of some of the past developments should do the following: (1) place the present analytical method in better perspective, (2) permit a more complete understanding of the interrelated variables involved, and (3) permit an evaluation of the methods' capabilities and limitations. For these reasons, the evolution of the basic FDA multiresidue method will be traced chronologically. References will be almost entirely limited to investigations conducted by FDA's staff scientists. However, other literature citations in these reports give some of the scientific precedent for the work and should provide a broad view of developments in multiresidue pesticide analytical methodology. An appendix contains references to supporting data grouped as to subject area, *i.e.*, gas chromatography, extraction, etc.; a flow diagram of the analytical scheme; and a list of pesticide chemicals recovered by the method.

The FDA "Pesticide Analytical Manual," Volume I,[2] January 1968, July 1969, and June 1970 revisions (PAM I), contains the methods and supporting data developed by FDA scientists for the analysis of foods for multiple residues of pesticide chemicals. Although more than one analytical method is included in this manual, only the widely applicable and routinely used scheme which is often referred to as the "Mills method" will be considered in this review. This method has evolved over a period of ten to 12 years with continual extension and improvement. We at FDA feel that development of a generally useful and reliable multiresidue method must be founded on a sound, basic approach which can be extended and refined as necessary and as new information or technology is developed. The revision procedure of FDA's PAM I is designed to keep pace with refinements in multiresidue methodology.

PAM I, Chapter 2, Sections 211 and 212, give the multiresidue methods application to fatty and nonfatty foods, respectively, for residues of organochlorine pesticides. Analysis for a limited number of organophosphorus pesticides by the same method is covered in Sections 231 and 232 for fatty and nonfatty foods, respectively. Determination by gas chromatography (GLC) using electron capture, thermionic, microcoulometric, or a combination of electron capture and thermionic detectors is covered in Chapter 3. Extensive relative retention time and detector response data are given in this chapter for three GLC columns and electron capture, microcoulometric, and thermionic detectors. Thin-layer chromatographic (TLC) procedures and pesticide migration data are given in Chapter 4, Sections 410, 411, and 412.

[2] FDA Pesticide Analytical Manual: Volume I contains methods which determine multiple pesticide residues, Volume II contains methods for individual pesticides. Obtain from U. S. Food and Drug Administration, Office of Associate Commissioner for Compliance, 200 C St., S.W., Washington, D. C. 20204.

III. Evolution of the FDA multiresidue method

"Detection and Semiquantitative Estimation of Chlorinated Organic Pesticide Residues in Foods by Paper Chromatography" by MILLS (1959) provided a general procedure to analyze nonfatty or fatty foods for multiple residues of up to 13 organochlorine pesticides. Pesticides were extracted from fresh fruits and vegetables with a water miscible-water immiscible solvent pair as suggested by GUNTHER and BLINN (1955). KLEIN *et al.* (1959) proved the effectiveness of this extractant for removing weathered residues resulting from foliar spraying in a classical study with radiolabelled methoxychlor. KLEIN (1958) had previously demonstrated the ineffectiveness of extraction with a water-immiscible solvent alone or extraction by surface stripping. Pesticides were extracted from food products containing fats by classical fat extraction procedures. Residues of the organochlorine pesticides were assumed to be in extracted fat. Pesticides were extracted from fats by partitioning a hydrocarbon solution of the fat with acetonitrile as suggested earlier by JONES and RIDDICK (1952). Florisil (1968), a synthetic magnesium silicate patented by FITZ-SIMMONS (1946),[3] was adapted to a column chromatographic procedure for removing co-extracted sample material. Eleven pesticides were eluted with 200 ml. of six percent ethyl ether/petroleum ether. Saponification and chromatography on magnesium oxide/Celite or acid Celite were included as alternate or optional cleanup techniques. Detection and semiquantitative estimation were made by paper chomatography utilizing techniques developed by MITCHELL (1957). MITCHELL's (1957) "Ascending Paper Chromatography—A Way to Do It" is a comprehensive source of information on determination of organochlorine pesticides by paper chromatography. MILLS (1961) reported on a collaborative study of this method for six common organochlorine pesticides in milk. The method, using paper chromatographic determination, is included in "Official Methods of Analysis" of the *Association of Official Analytical Chemists* (AOAC) (HORWITZ 1970) as a qualitative test for DDT, TDE, heptachlor epoxide, BHC, and lindane in dairy products.

BURKE and JOHNSON (1961 and 1962) reported on the application of gas chromatography with COULSON's (1960) halogen-specific microcoulometric detection system to samples prepared by the MILLS (1959) method. Gas chromatographic conditions suitable for routine use in multiresidue analysis were established and relative retention times for a large number of pesticide chemicals were determined.

Following the report of LOVELOCK and LIPSKY (1960) on detection of certain GLC effluents by the electron capture phenomenon, BOGGS

[3] Manufactured by the Floridin Company, Berkeley Springs, W. Va. 25411; also, 300 Penn Center Blvd., Pittsburgh, Pa. 15222.

and LIPSCOMB (1961) and LIPSCOMB (1963) introduced electron capture GLC into FDA's pesticide residue analytical program in the Buffalo, New York laboratory. WELLS (1962) and GREENSPAN *et al.* (1962) described the use of this detection technique in certain analyses for organochlorine pesticides in other FDA district laboratories. WATTS and KLEIN (1962) further studied the use of electron capture GLC for the determination of organochlorine pesticide residues.

JOHNSON (1962) developed an extraction procedure using acetonitrile instead of the hexane-isopropanol extractant in the MILLS (1959) method for fruits and vegetables. Acetonitrile extracted quantities of weathered residues comparable to hexane-isopropanol and had the advantages of removing less coloring matter than hexane-isopropanol and of being a single solvent extractant. MODDES and COOK (1959) had earlier utilized acetonitrile for extracting parathion, and STORHERR and BURKE (1961) had recognized its potential as a good solvent for pesticides. JOHNSON (1962) also extended the method to dieldrin and endrin by an elution of the Florisil column with 200 ml. of 15 percent ethyl ether/petroleum ether following the first elution with 200 ml. of six percent ethyl ether/petroleum ether. BURKE and MILLS (1963) showed that endosulfan and tetradifon could be quantitatively recovered by eluting the Florisil column with an eluant of 30 percent ethyl ether/petroleum ether.

The details of a TLC scheme for detection of organochlorine pesticides were reported by KOVACS (1963). Aluminum oxide-coated plates were developed with either *n*-heptane or two percent acetone/*n*-heptane and pesticides were visualized by spraying with a silver nitrate solution described by MITCHELL (1956). This procedure is the basis of FDA's current TLC of organochlorine pesticides.

MILLS *et al.* (1963) combined and refined portions of previously reported procedures to provide a rapid and routinely applicable method for quantitative analysis of residues of organochlorine pesticides in nonfatty foods. Twenty-one pesticide chemicals were known to be recovered. Residues were extracted by a single blending with acetonitrile; the filtered extract represented an aliquot of the total volume of acetonitrile extractant plus sample moisture. The extract was cleaned up by Florisil column chromatography with successive elutions using 200 ml. each of six percent and 15 percent ethyl ether/petroleum ether. Pesticides were determined by microcoulometric GLC. Recovery data were presented for five pesticides on seven vegetables, two fruits, and two canned soups. The MILLS *et al.* (1963) method of extraction and cleanup is the basic multiresidue methodology now used by FDA for nonfatty foods. It has been extended to many additional sample types and pesticides and continues to be the basis for further extensions and refinements.

SHUMAN (1963) demonstrated that Anakrom ABS was superior to the previously used Chromosorb P for GLC of pesticides sensitive to interaction with the solid support. BURKE and GIUFFRIDA (1964) investigated GLC with electron capture detection for the purpose of defining, improving, and standardizing its use in analysis for organochlorine residues. Instrumentation and operating parameters were selected for simplicity and maximum performance in routine residue analysis. Electron capture detection sensitivity was standardized, and relative retention time and response data, with ten percent DC-200 liquid phase on Anakrom ABS solid support, were determined for a large number of pesticide chemicals. The necessity for sample cleanup and use of high purity reagents was shown. These are the basic GLC system and parameters currently used in FDA laboratories.

GIUFFRIDA (1964) reported the development of a flame detector rendered highly sensitive and therefore selective to organophosphorus compounds by placement of an alkali salt in the flame. A new era in methods development and in the analysis for residues of organophosphorus pesticide chemicals began with this detector. GIUFFRIDA's report indicated that the MILLS *et al.* (1963) method may also be applicable to certain organophosphorus pesticides. NELSON (1964) showed that certain parent organophosphorus pesticides could be recovered through the extraction and solvent partitioning steps of the MILLS *et al.* (1963) method. Several FDA investigators were involved in demonstrating that parathion, methyl parathion, diazinon, ethion, carbophenothion, and ronnel could be quantitatively recovered by this method in addition to many organochlorine pesticides.

GLC relative retention time and response data for a large number of organochlorine and organosulfur pesticide chemicals using improved microcoulometric instrumentation and ten percent DC-200 liquid phase were reported by BURKE and HOLSWADE (1964).

KOVACS (1964) reported a TLC procedure suitable for thiophosphorus pesticides. Aluminum oxide-coated plates with an immobile phase of dimethyl formamide were developed with methylcyclohexane. Pesticides were visualized with tetrabromophenolphthalein ethyl ester, silver nitrate, and citric acid.

A successful collaborative study of the MILLS (1959) method, utilizing electron capture GLC (BURKE and GIUFFRIDA 1964) and TLC (KOVACS 1963) for quantitative and qualitative determination, respectively, of dieldrin and heptachlor epoxide in milk and butterfat was reported by JOHNSON (1965). Prior to this study, investigation by MILLS and BURKE (1964) and FURMAN and FEHRINGER (1967) resulted in certain modifications of the method to improve its quantitative capabilities. This method is included in "Official Methods of Analysis" of the AOAC (HORWITZ 1970) for residues of dieldrin and heptachlor epoxide in dairy products and vegetable oils and is currently used by FDA for analysis of fatty foods.

GIUFFRIDA and IVES (1965) showed that Gas Chrom Q was an improvement over Anakrom ABS as a GLC solid support for chromatography of sensitive compounds. BURKE and HOLSWADE (1966) reported GLC relative retention time and response data for a large number of pesticide chemicals on a column of equal parts 15 percent QF-1 and ten percent DC-200 with Gas Chrom Q solid support. This column gives retention times different from those obtained with a DC-200 liquid phase and is used routinely along with ten percent DC-200 in FDA's residue analyses program.

Successful collaborative studies of the MILLS *et al.* (1963) method, using electron capture GLC (BURKE and GIUFFRIDA 1964) and TLC (KOVACS 1963) for quantitative and qualitative analysis, respectively, for aldrin, *p,p'*-DDE, methoxychlor, lindane, heptachlor, *p,p'*-TDE, BHC, *p,p'*-DDT, and endrin in five fruits and vegetables were reported by KRAUSE (1966), GAUL (1966), and DAVIDSON (1966). This method is included in "Official Methods of Analysis" of the AOAC (HORWITZ 1970) for residues of the above compounds in a variety of common fruits and vegetables. The method is in current use by FDA for a large number of nonfatty type foods.

SAWYER (1966) demonstrated that acetonitrile was a feasible and practical solvent for the extraction of organochlorine residues from raw eggs and that previously reported GLC interferences (ONLEY and MILLS 1962) did not often occur with this method.

BURKE and PORTER (1966) used kale field-sprayed with TDE to show that the MILLS *et al.* (1963) acetonitrile extraction quantitatively extracts weathered residues of a foliar-applied chlorinated pesticide. The effectiveness of other extraction techniques was also reported. Investigations with residues of foliar applied methoxychlor, endosulfan, and aldrin (dieldrin) confirmed and extended the 1966 results (BURKE *et al.* 1970).

MALONE (1966) found that water washing of ethyl ether to remove peroxides and ethanol prior to re-establishing ethanol content was not completely effective for these purposes and was unnecessary as a routine treatment of ethyl ether used for preparation of Florisil eluants. WIENCKE and BURKE (1969 a) and GAUL (1969 a) showed that the presence of peroxides in Florisil eluants could result in losses of some organophosphorus pesticides, particularly carbophenothion, during evaporation of column eluates. Calcination conditions used in the manufacture of Florisil were studied by BURKE and MALONE (1966). Conditions of calcination and a column chromatography elution test with certain chlorinated pesticides were established by the manufacturer to attain a more reliable Florisil for use in FDA's multiresidue methods.

LAWRENCE (1966) and LAWRENCE and ROACH (1967) reported GLC relative retention time and response data for an additional number of pesticide chemicals on both ten percent DC-200 and 15 percent QF-1/ten percent DC-200 liquid phases.

WESSEL (1967) reported on a successful collaborative study of the MILLS *et al.* (1963) method for parathion, methyl parathion, ethion, carbophenothion, diazinon, ronnel, and malathion in lettuce and apples. Malathion was eluted from Florisil with a third eluant, 50 percent ethyl ether/petroleum ether, as previously found by FDA investigators. Determination was by dual electron capture-flame thermionic detectors after gas chromatography on ten percent DC-200 (BURKE and GIUFFRIDA 1964, GIUFFRIDA 1964). TLC (KOVACS 1964) was used to check pesticide identity. This method is included in "Official Methods of Analysis" of the AOAC (HORWITZ 1970) for the above pesticides and crops and is in routine use by FDA laboratories. A TLC system for improved separation and detection of ronnel, ethion, and carbophenothion was developed by GARDNER (1967).

BERTUZZI *et al.* (1967) demonstrated that acetonitrile alone was not satisfactory for extracting chlorinated pesticides from dehydrated samples. However, 35 percent water/acetonitrile was shown to be an effective extractant which could be conveniently used in the MILLS *et al.* (1963) method for dry samples. Using kale with residues of TDE, BURKE and PORTER (1967) showed that there was a direct relationship between sample moisture content and extraction efficiency of acetonitrile. The quantity of pesticide extracted decreased as sample moisture decreased below about 70 percent. The combination of 35 percent water/acetonitrile proved to be a good extractant for samples with reduced moisture content. A similar study (BURKE *et al.* 1970) with methoxychlor and endosulfan residues verified the previous work on the relationship of sample moisture and extraction efficiency.

PORTER *et al.* (1967) found that the acetonitrile dilution operation in the MILLS *et al.* (1963) method could be a source of low results and suggested a means for improving partitioning of organochlorine pesticides from aqueous acetonitrile to petroleum ether. GAUL (1969 b) and PORTER *et al.* (1969) studied the behavior of several organochlorine and organophosphorus pesticides in this operation.

WELLS (1967) reported on recovery experiments with the FDA multiresidue methods for nonfatty and fatty type samples. Eighteen laboratories conducted experiments involving 65 pesticide chemicals and a wide variety of foods and animal feeds. Samples comprised 67 nonfatty types and 17 foods containing over two percent fat. Recoveries for 32 compounds were within the acceptable range of 80 to 110 percent. Recoveries for 20 compounds were outside these limits and 13 compounds could not be recovered by the method. These data were the basis of BURKE's (1967) recommendation to the *Association of Official Analytical Chemists* that the official method be extended to 13 additional crops and to vegetable oils. Only those recovery data for pesticides previously covered by the official method were considered for the extension recommended.

MILLS (1968) reported a procedure to determine the adsorptive capacity of Florisil. Based on adsorptions of lauric acid from hexane solution, the amount of adsorbent used in the column chromatography cleanup can be determined empirically to adjust for some differences among the manufacturer's lots.

It was shown by Porter and Burke (1968) that the Mills *et al.* (1963) acetonitrile extraction quantitatively removed parathion and diazinon residues resulting from foliar spraying.

Thomas *et al.* (1968) reported TLC migration data for 40 pesticides with three solvent systems often employed by FDA to obtain qualitative information on organochlorine pesticides. A two-dimensional TLC system for the separation and detection of 25 parent organophosphorus pesticides was developed by Gardner (1970).

Recovery data for nine organochlorine pesticides in 21 crops were presented by Burke (1968), enabling extension of the *Association of Official Analytical Chemists* method (Horwitz 1970) to 15 additional crops, making a total of 32 fruits and vegetables, dairy products, and vegetable oils for which the method is considered official.

Ives and Giuffrida (1968) found Chromosorb W HP to be superior to Gas Chrom Q as a GLC solid support for chromatographing a range of diverse pesticide types. Chromosorb W HP was more inert to reactive compounds.

Additional information showing the effectiveness of the 35 percent water/acetonitrile extractant for dry products was reported by Wilderman and Shuman (1968); DDT residues were completely extracted from hay.

Wessel (1968) conducted an interlaboratory study of three modes of combining electron capture and flame thermionic GLC detection. It was shown that any one of the three techniques was suitable for use in quantitative analysis when assembly was proper and certain operating parameters adhered to. A series arrangement with the complete column effluent passing through the electron capture detector to the thermionic detector is used by most laboratories.

Wessel (1969) demonstrated a favorable comparison of the Mills *et al.* (1963) acetonitrile extraction to a chloroform-methanol Soxhlet extraction (Mumma *et al.* 1966) for the quantitative extraction of residues of six organochlorine pesticides from eggs.

Recently, several investigators have developed procedures which "branch out" from the Mills *et al.* (1963) method for the purpose of analyzing for compounds or groups of compounds not included in the basic multiresidue method. This is part of a continuing effort toward a more unified and complete scheme of analysis for residues of pesticide chemicals of many types. Storherr (1970) showed that an aliquot of the acetonitrile extract of fruits and vegetables could be cleaned up by column chromatography on charcoal (Watts *et al.* 1969) for analysis for a number of organophosphorus compounds not recovered by the Mills *et al.* (1963) method. Pomerantz *et al.* (1970) described a procedure for captan, folpet, and difolatan; these fungicides were extracted from the diluted acetonitrile extract with methylene chloride/petroleum ether and were determined by gas chromatography after cleanup by Florisil chromatography.

PORTER *et al.* (1969) then showed that carbaryl (Sevin) was quantitatively extracted from fresh fruits and vegetables with acetonitrile and perfected a method of analysis for carbaryl in an aliquot of the extract normally used for multiresidue analysis. Polarography, TLC, or spectrophotometry may be used for the determination.

ONLEY and YIP (1969) described a method for the determination of eight urea herbicides, some of their metabolites, and 3-aminotriazole. These compounds were extracted from the normally discarded, diluted acetonitrile with methylene chloride/carbon tetrachloride after the usual partitioning of the nonpolar organochlorine and organophosphorus pesticides into petroleum ether. A combination of column chromatographic techniques was used for cleanup. The ureas were determined by TLC and 3-aminotriazole was measured spectrophotometrically.

BURKE (1969) reported Florisil elution, GLC, and TLC characteristics of two dieldrin photoconversion products and showed that these compounds would be determined by the FDA multiresidue methods for nonfatty and fatty samples.

ARMOUR and BURKE (1969) showed that the FDA multiresidue methods for nonfatty and fatty samples would completely recover and determine polychlorinated biphenyls[4] and gave GLC, TLC, and Florisil elution characteristics for this group of compounds. ARMOUR and BURKE (1970) developed a procedure, using column chromatography on silicic acid, to separate polychlorinated biphenyls from DDT, its analogs, and other organochlorine pesticides. The separation is performed on the six percent ethyl ether/petroleum ether eluate from the Florisil column. Organochlorine pesticides and polychlorinated biphenyls can be determined without cross interference.

A comparison of ten methods for the analysis of milk for residues of organochlorine pesticides was made by LAWRENCE and BURKE (1969). Results on a residue present in the milk and an added pesticide indicated the MILLS method (MILLS 1959, JOHNSON 1965) among those methods giving quantitative results and suitable cleanup for multiresidue analysis.

Simplified and improved GLC equipment for use with microcoulometric detection was described by GIUFFRIDA and IVES (1969). Effective detection sensitivity of the improved system is 1/10 to 1/15 of the electron capture system's sensitivity permitting microcoulometric examination of many residues initially indicated by electron capture GLC.

CARR (1970) conducted a collaborative study of the MILLS method

---

[4] Editor's note: See REYNOLDS, L. M.: Pesticide residue analysis in the presence of polychlorobiphenyls (PCB's). Residue Reviews, this volume.

(Mills 1959, Johnson 1965) for α-BHC, lindane, *p,p'*-DDE, *p,p'*-TDE, *o,p'*-DDT, *p,p'*-DDT, and methoxychlor in butterfat. Good results were obtained and these compounds were included in the official AOAC method for pesticide residues in dairy products: see "Official Methods of Analysis" of the AOAC (Horwitz 1970).

Laski (1969) described a GLC column with 15 percent QF-1 liquid phase for separation of α, β, γ, and δ isomers of BHC. Miller (1970) separated these isomers of BHC, using three percent OV-17 liquid phase, and presented relative retention time data for a number of organochlorine pesticide chemicals. This separation is not complete on the DC-200 and QF-1/DC-200 liquid phases routinely used for multiresidue analyses.

Porter and Burke (1969) described changes found necessary in the extraction-filtration process of the Mills *et al.* (1963) method to assure that the filtrate was a representative fraction of the extract. These changes were made to eliminate the possibility of high results occasionally obtained with samples of high sugar content.

Porter *et al.* (1970) studied and refined a procedure to extract organochlorine pesticide residues from fish, meat, and poultry tissue. Blending with petroleum ether was shown to make a good extraction of metabolically incorporated residues of organochlorine pesticides from fish of various fat contents, beef, pork, and poultry. Cleanup involves partitioning between petroleum ether and acetonitrile and Florisil column chromatography (Mills 1959, Johnson 1965, Carr 1970). An additional partitioning of acetonitrile with petroleum ether (Johnson and Throm 1964, Moya 1967) reduces the weight of fatty extractives by about a half with little effect on recovery of pesticides. This method as applied to fish is presently under collaborative study in a number of laboratories (Carr 1970).

Burke *et al.* (1970) demonstrated that the acetonitrile extraction of Mills *et al.* (1963) accounted for only 50 to 75 percent of dieldrin residues present in potato tubers as a result of adsorption from dieldrin-treated soil. Potatoes were grown in both the greenhouse and the field. Similar results were obtained with the 35 percent water/acetonitrile extractant (Bertuzzi *et al.* 1967) when applied to the potatoes after they had been dehydrated.

Pardue (1970) found that recoveries were in excess of 80 percent for 13 of 60 additional organophosphorus pesticides and oxygen analogs studied in the method of Mills *et al.* (1963). Two of the compounds recovered required elution of the Florisil column with 50 percent ethyl ether/petroleum ether. Recoveries of 33 to 76 percent were obtained for eight other compounds.

IV. Confirmation of residue identity

A definite procedure or scheme for confirming the identity of a given pesticide residue has not been established. Thin-layer chromatography, and GLC on columns that give different retention patterns, and the p-value systems of BEROZA and BOWMAN (1965) and BOWMAN and BEROZA (1965)[5] are probably the most universally applicable confirmatory techniques. Derivatization of the residue in question followed by GLC of the derivative seems an excellent approach and several procedures have appeared in the literature recently. KLEIN and WATTS (1964), HAMMENCE *et al.* (1965), SANS (1967), DUFFY and WONG (1967), OSADCHUK and WANLESS (1968), COCHRANE and CHAU (1968), CHAU and COCHRANE (1969 a and b), CHAU (1969), and WIENCKE and BURKE (1969 b) have developed procedures for forming characteristic chemical derivatives for a number of organochlorine pesticides.

BANKS and BILLS (1968) used ultraviolet irradiation to convert organochlorine insecticides to gas chromatographable degradation products. MILLER and WELLS (1969) used an alkaline pre-column to change certain compounds within the GLC system.

One or more tests in addition to the initial analyses are usually necessary to develop information sufficient to identify the residue. The appropriate tests must be selected by the analyst. Therefore it is in residue identification that the training, experience, and expertise of the analyst play the major role in obtaining an accurate result. The importance of the highly qualified analyst or laboratory in obtaining a reliable residue analysis with any method cannot be overemphasized.

V. Limitations

The primary limitations of the FDA multiresidue methodology are those that are common to all multiresidue analytical methods. Many pesticide chemicals are not analyzed for and not all sample types can be handled in a routine fashion. All the variables in procedures for oilseeds, dietary composites, feeds, and grains have not been thoroughly studied and current approaches may require revision. Complete extraction may not be obtained for residues absorbed from the soil into the plant root or translocated from the root to aerial parts of the plant. Cleanup is not always satisfactory for all samples, particularly fatty types, forcing the analyst to resort to various supplemental procedures. As mentioned above, there is no complete scheme for confirming the identity of a large number of pesticide chemicals. These areas will continue to receive attention by FDA's method development laboratories.

---

[5] Editor's note: BEROZA, M., M. N. INSCOE, and M. C. BOWMAN: Distribution of pesticides in immiscible binary solvent systems for cleanup and identification and its application in the extraction of pesticides from milk. Residue Reviews **30**, 1–61 (1969).

VI. Appendix

a) Reference to data in support of methodology

1. Extraction

Method

Fatty samples:	PAM[6] 211.101–211.14
	AOAC[7] 29.010–29.011
Nonfatty samples:	PAM[6] 212.101–212.13a(2)
	AOAC[7] 29.009

Supporting data (chronological order)

Klein, A. K.: Report on extraction procedures for chloroorganic pesticides. J. Assoc. Official Agr. Chemists **41**, 551 (1958).

———, E. P. Laug, and J. D. Sheehan, Jr.: Report on extraction procedures for chloroorganic insecticides. J. Assoc. Official Agr. Chemists **42**, 539 (1959).

Mills, P. A., J. H. Onley, and R. A. Gaither: Rapid method for chlorinated pesticide residues in nonfatty foods. J. Assoc. Official Agr. Chemists **46**, 186 (1963).

Sawyer, A. D.: Study of extraction procedures for chlorinated organic pesticides in fresh eggs. J. Assoc. Official Anal. Chemists **49**, 643 (1966).

Burke, J. A., and M. L. Porter: A Study of the effectiveness of some extraction procedures for pesticide residues in vegetables. J. Assoc. Official Anal. Chemists **49**, 1157 (1966).

Bertuzzi, P. F., L. Kamps, C. I. Miles, and J. A. Burke: Extraction of chlorinated pesticide residues from nonfatty samples of low moisture content. J. Assoc. Official Anal. Chemists **50**, 623 (1967).

Burke, J. A., and M. L. Porter: Note on the effect of sample moisture content on extraction of TDE from kale. J. Assoc. Official Anal. Chemists **50**, 1260 (1967).

Porter, M. L., and J. A. Burke: A preliminary study of the extraction of parathion and diazinon by the Mills, Onley, Gaither acetonitrile procedure. J. Assoc. Official Anal. Chemists **51**, 63 (1968).

Wilderman, M., and H. Shuman: Extraction procedures for DDT residues in firm fruits and in hay: Colorimetric method. J. Assoc. Official Anal. Chemists **51**, 892 (1968).

Lawrence, J. H., and J. A. Burke: Comparison of ten methods for the analysis of milk for residues of chlorinated pesticides. J. Assoc. Official Anal. Chemists **52**, 817 (1969).

Wessel, J. R.: Comparison of four extraction and cleanup procedures for chlorinated pesticide residues in eggs. J. Assoc. Official Anal. Chemists **52**, 172 (1969).

Porter, M. L., R. Gajan, and J. A. Burke: Acetonitrile extraction and determination of carbaryl in fruits and vegetables. J. Assoc. Official Anal. Chemists **52**, 177 (1969).

———, and J. A. Burke: Modification of the AOAC method for multiple pesticide

[6] PAM = FDA Pesticide Analytical Manual, Vol. I.

[7] AOAC = Association of Official Analytical Chemists.

residues in nonfatty foods for use on high sugar samples. J. Assoc. Official Anal. Chemists **52**, 1280 (1969).

———, S. J. V. YOUNG, and J. A. BURKE: Analysis of fish, animal, and poultry tissue for chlorinated pesticide residues. Submitted to J. Assoc. Official Anal. Chemists.

BURKE, J. A., M. L. PORTER, and S. J. V. YOUNG: Extraction of foliar and root adsorbed pesticide residues from fresh and dehydrated nonfatty samples by the Mills, Onley, Gaither acetonitrile procedure. Submitted to J. Assoc. Official Anal. Chemists.

2. Cleanup

Method

Fatty samples:	PAM[6] 211.14–211.16d
	AOAC[7] 29.011, 29.014–29.016
Nonfatty samples:	PAM[5] 211.15–211.16d
	AOAC[6] 29.014–29.016

Supporting data (chronological order)

MILLS, P. A.: Detection and semiquantitative estimation of chlorinated organic pesticide residues in foods by paper chromatography. J. Assoc. Official Agr. Chemists **42**, 734 (1959).

JOHNSON, L.: Separation of dieldrin and endrin from other chlorinated pesticide residues. J. Assoc. Official Agr. Chemists **45**, 363 (1961).

BURKE, J. A., and P. A. MILLS: Microcoulometric gas chromatographic determination of Thiodan and Tedion in green vegetables. J. Assoc. Official Agr. Chemists **46**, 177 (1963).

MILLS, P. A., J. H. ONLEY, and R. A. GAITHER: Rapid method for chlorinated pesticide residues in nonfatty foods. J. Assoc. Official Agr. Chemists **46**, 186 (1963).

BURKE, J. A., and B. MALONE: Effects of calcination temperature and time on retentive properties of Florisil used for pesticide residue analysis. J. Assoc. Official Anal. Chemists **49**, 1003 (1966).

MALONE, B.: Preparation of ethyl ether for use in Florisil column chromatography eluants. Private communication, *U. S. Food and Drug Administration* (November 1966).

BURKE, J. A., P. A. MILLS, and D. C. BOSTWICK: Experiments with evaporation of solutions of chlorinated pesticides. J. Assoc. Official Anal. Chemists **49**, 999 (1966).

PORTER, M. L., J. A. BURKE, and P. F. BERTUZZI: Acetonitrile dilution: A possible source of error in the "Rapid method for chlorinated pesticide residues in nonfatty foods." J. Assoc. Official Anal. Chemists **50**, 644 (1967).

FURMAN, W. B., and N. V. FEHRINGER: Study of modifications of the Mills procedure for chlorinated pesticides in dairy products. J. Assoc. Official Anal. Chemists **50**, 903 (1967).

MILLS, P. A.: Variation of Florisil activity: Simple method for measuring adsorbent capacity and its use in standardizing Florisil columns. J. Assoc. Official Anal. Chemists **51**, 29 (1968).

PORTER, M. L., S. J. V. YOUNG, and J. A. BURKE: Analysis of fish, animal, and poultry tissue for chlorinated pesticide residues. Submitted to J. Assoc. Official Anal. Chemists.

3. Gas chromatography

Method

Microcoulometric detection:	PAM[6] 312–312.71
Electron capture detection:	PAM[6] 311–311.6C
	AOAC[7] 29.008(a–c), 29.017 (a–b)
Potassium chloride thermionic detector:	PAM[6] 313–313.5
	AOAC[7] 29.008 (d–g, k)
Dual electron capture/potassium chloride thermionic detector:	PAM[6] 321.5
	AOAC[7] 29.017(c)

Supporting data (chronological order)

Burke, J. A., and L. Johnson: Investigations in the use of the microcoulometric gas chromatograph for pesticide residue analysis. J. Assoc. Official Agr. Chemists **45**, 348 (1962).

Giuffrida, L.: A flame ionization detector highly selective and sensitive to phosphorus—A sodium thermionic detector. J. Assoc. Official Agr. Chemists **47**, 293 (1964).

Burke, J. A., and L. Giuffrida: Investigations of electron capture gas chromatography for the analysis of multiple chlorinated pesticide residues in vegetables. J. Assoc. Official Agr. Chemists **47**, 326 (1964).

———, and W. Holswade: Gas chromatography with microcoulometric detection for pesticide residue analysis. J. Assoc. Official Agr. Chemists **47**, 845 (1964).

Giuffrida, L., and N. F. Ives: Investigation of two gas chromatographic techniques for the determination of organophosphate pesticide residues. J. Assoc. Official Agr. Chemists **47**, 1112 (1964).

Burke, J. A.: Gas chromatography for pesticide residue analysis: Some practical aspects. J. Assoc. Official Agr. Chemists **48**, 1037 (1965).

Giuffrida, L., N. F. Ives, and D. C. Bostwick: Gas chromatography of pesticides—Improvements in the use of special ionization detection systems. J. Assoc. Official Anal. Chemists **49**, 8 (1966).

Burke, J. A., and W. Holswade: A gas chromatographic column for pesticide residue analysis: Retention and response data. J. Assoc. Official Anal. Chemists **49**, 374 (1966).

Ives, N. F., and L. Giuffrida: Investigation of thermionic detector responses for the gas chromatography of P, N, As, and Cl organic compounds. J. Assoc. Official Anal. Chemists **50**, 1 (1967).

Giuffrida, L., and N. F. Ives: Microcoulometric gas chromatography: Improvements in instrumentation. J. Assoc. Official Anal. Chemists **52**, 541 (1969).

4. Thin-layer chromatography

Method

PAM[6] 410.1–413.2
AOAC[7] 29.018–29.026

Supporting data and other procedures (chronological order)

KOVACS, M. F.: Thin layer chromatography for chlorinated pesticide residue analysis. J. Assoc. Official Agr. Chemists **46**, 884 (1963).

——— Thin layer chromatography for organothiophosphate pesticide residue determination. J. Assoc. Official Agr. Chemists **47**, 1097 (1964).

FEHRINGER, N. V., and OGGER, J. D.: Thin layer chromatography identification of pesticides using dip-coated presensitized microscope slides. J. Chromatog. **25**, 95 (1966).

KOVACS, M. F.: Rapid detection of chlorinated pesticide residues by an improved TLC technique: 3-¼″ × 4″ micro slides. J. Assoc. Official Anal. Chemists **49**, 365 (1966).

GARDNER, A.: The separation and detection of ronnel, ethion, and carbophenothion on Silica Gel G thin layer chromatographic plates. Private communication *U. S. Food and Drug Administration* (August 1967).

THOMAS, E., J. A. BURKE, and J. LAWRENCE: Thin layer chromatography: Relative migration data (R_{TDE}) of chlorinated pesticides. J. Chromatog. **35**, 119 (1968).

GARDNER, A.: A two dimensional TLC procedure for confirming fifteen parent organophosphate pesticides. Private communication *U. S. Food and Drug Administration* (September 1968).

5. Collaborative studies and validation data (chronological order)

MILLS, P. A.: Collaborative study of certain chlorinated organic pesticides in dairy products. J. Assoc. Official Agr. Chemists **44**, 171 (1961).

JOHNSON, L.: Collaborative study of a method for multiple chlorinated pesticide residues in fatty foods. J. Assoc. Official Agr. Chemists **48**, 668 (1965).

KRAUSE, R.: Collaborative study of a method for multiple chlorinated pesticide residues in nonfatty vegetables. J. Assoc. Official Anal. Chemists **49**, 460 (1966).

GAUL, J.: Collaborative study of a method for multiple chlorinated pesticide residues in leafy and cole-type vegetables. J. Assoc. Official Anal. Chemists **49**, 463 (1966).

DAVIDSON, A.: Collaborative study of a rapid method for multiple chlorinated pesticide residues in small fruits. J. Assoc. Official Anal. Chemists **49**, 468 (1966).

WESSEL, J. R.: Collaborative study of a method for multiple organophosphorus pesticide residues in nonfatty foods. J. Assoc. Official Anal. Chemists **50**, 430 (1967).

WELLS, G.: Validation study of a method for pesticide residues in foods and animal feeds. J. Assoc. Official Anal. Chemists **50**, 1205 (1967).

BURKE, J. A.: Report on chlorinated pesticides. J. Assoc. Official Anal. Chemists **51**, 311 (1968).

WESSEL, J. R.: Collaborative study of three gas chromatographic dual detection systems for analysis of multiple chlorinated and organophosphorus pesticides. J. Assoc. Official Anal. Chemists **51**, 666 (1968).

CARR, R. L.: Collaborative study of a method for multiple chlorinated pesticide residues in fatty foods. J. Assoc. Official Anal. Chemists **53**, 152 (1970).

——— Collaborative study of a method for analysis of fish for multiple residues of chlorinated pesticides. Submitted to J. Assoc. Official Anal. Chemists.

GARDNER, A.: Confirmation of organophosphorus pesticide residues at nanogram levels by two dimensional thin layer chromatography. Submitted to J. Assoc. Official Anal. Chemists.

b) *Outline of method*

CHLORINATED (NONIONIC) AND CERTAIN ORGANOSPHOSPHATE RESIDUES

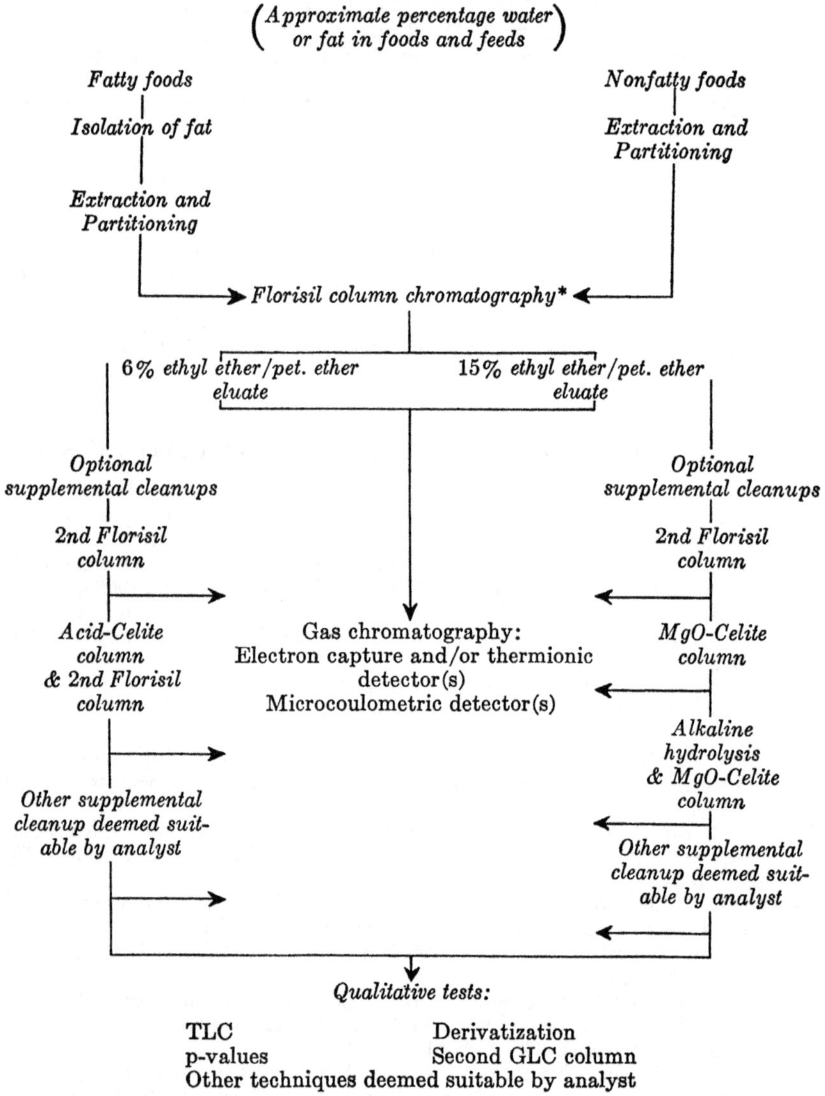

* Third elution with 50 percent ethyl ether/pet. ether made with certain sample types or certain residues.

c) *Pesticides recovered by method*

1. Fatty foods

Table I. *Pesticides recovered through method for fatty foods*[a,b]

Eluate from Florisil	
6% eluate	15% eluate
aldrin	bulan (<80%)
BHC (α,β,γ,δ)[c]	*n*-butyl ester 2,4-D (<80%)
carbophenothion (<80%)	*n*-butyl ester 2,4,5-T (<80%)
chlorbenside	butoxy ethanol ester 2,4-D (<80%)
chlordane (technical insecticide)	butoxy ethanol ester 2,4,5-T (<80%)
p,p'-DDT[c]	chlordecone (<80%)
o,p'-DDT[c]	chlorobenzilate (<80%)
p,p'-DDE[c]	Chlorothion®
o,p'-DDE	CIPC
dicofol (6% and 15%) (<80%)	Dacthal® (<80%)
ethion	diazinon
heptachlor	dicofol (6% and 15%) (<80%)
heptachlor epoxide[c]	dieldrin[c]
hexachlorobenzene (<80%)	Dilan® (<80%)
isobenzan	Dyrene® (<80%)
lindane[c]	endosulfan I
Merphos® (6% and 15%)	endrin
methoxychlor[c]	EPN
octachlor epoxide	ethyl hexyl ester 2,4-D
Perthane®	folpet (<80%)
Perthane® olefin	isobutyl ester 2,4-D
phorate (<80%)	iso-octyl ester 2,4,5-T
polychlorinated biphenyls	iso-octyl ester 2,4-D (<80%)
quintozene	isopropyl ester 2,4,5-T
ronnel	isopropyl ester 2,4-D
Strobane®	Merphos® (6% and 15%)
o,p'-TDE	methyl parathion
p,p'-TDE[c]	ovex
p,p'-TDE olefin	parathion
tecnazene	prolan (<80%)
tetraiodoethylene (<80%)	tetradifon
toxaphene	
trifluralin	
Vegadex®	

[a] Method described in "Official Methods of Analysis" of the AOAC. 11th Edition (1970); not including supplemental cleanups.

[b] Inclusion in list does not necessarily indicate that pesticide residue is likely to be the compound listed.

[c] Pesticides for which method is official AOAC.

2. Nonfatty foods

Table II. *Pesticides recovered through method for nonfatty foods*[a,b]

Eluate from Florisil	
6% eluate	15% eluate
aldrin[c]	bulan (<80%)
BHC (α,β,γ,δ)[c]	*n*-butyl ester 2,4-D
benefin	*n*-butyl ester 2,4,5-T (<80%)
bromophos-ethyl	butoxy ethanol ester 2,4-D (<80%)
carbophenothion[c]	butoxy ethanol ester 2,4,5-T (<80%)
chlorbenside	chlordecone
chlordane (technical insecticide)	chlorobenzilate (<80%)
p,p′-DDT[c]	chloropropylate
o,p′-DDT	Chlorothion®
p,p′-DDE[c]	CIPC
o,p-DDE	Dacthal®
dicofol (6% and 15%) (<80%)	diazinon[c]
DMPA	dialifor
Dursban®	dicofol (6% and 15%) (<80%)
Dyfonate®	dieldrin
ethion[c]	Dilan® (<80%)
heptachlor[c]	Dyrene®
heptachlor epoxide	endosulfan I
hexachlorobenzene	endrin[c]
isobenzan	EPN
isodrin	ethyl hexyl ester 2,4-D
lindane[c]	fenitrothion
Merphos® (6% and 15%)	folpet (<80%)
methoxychlor[c]	genite
Methyl Trithion®	isobutyl ester 2,4-D
mirex	iso-octyl ester 2,4,5-T
Nemacide®	iso-octyl ester 2,4-D
Perthane®	isopropyl ester 2,4,5-T
Perthane® olefin	isopropyl ester 2,4-D
phencapton	Merphos® (6% and 15%)
phorate (<80%)	methyl parathion[c]
polychlorinated biphenyls	ovex
quintozene	parathion[c]
ronnel[c]	prolan (<80%)
Strobane®	SD 7438
sulfotepp	tetradifon
o,p′-TDE	
p,p′-TDE[c]	
p,p′-TDE olefin	
tetraiodoethylene (<80%)	
tetrasul	
toxaphene	
trifluralin	
Vegadex®	

[a] Method described in "Official Methods of Analysis" of the AOAC. 11th Edition (1970); not including supplemental cleanups.

[b] Inclusion in list does not necessarily indicate that pesticide residue is likely to be the compound listed.

[c] Pesticides for which method is official AOAC.

3. Florisil elution; miscellaneous pesticides

Table III. *Florisil elution information for other pesticides*[a,b]

Pesticide	Comments
azinphosmethyl	Does not elute in 6% or 15%
n-butyl ester 2,4,5-T	Elutes in 15% and 30%
captan	Does not elute through 30%
chlordecone	Elutes in 15%, 20%, and 25%
chlorobenzilate	Elutes in 15% and 30%
coumaphos	Does not elute through 30%
DEF®	Elutes in 15% and 50%
dichlobenil	Elutes partially in 15%
dichloran	Elutes in 15% and 20%
dicofol	Elutes in 6%, or in both 6% and 15% inconsistently
Dilan®	Elutes in 15% and 20%
diuron	Requires 65% to elute
endosulfan II	Elutes in 20%, 25%, and 30%. Eluted by 30%
endrin alcohol	Elutes in 15% and 20%
endrin aldehyde	Elutes in 15% and 20%
endrin ketone (Delta ketone 153)	Elutes in 20% and 25%
hexachlorophene	Does not elute in 6%, 15%, or 50%
malathion[c]	Elutes in 50%
monuron	Requires 65% to elute
Merphos®	Elutes in 6% and/or 15% inconsistently
neburon	Does not elute through 30%
photodieldrin	90% elutes in 15%; eluted by 20%
phosalone	Elutes in 50%
Sulphenone®	Elutes in 20%, 25%, etc.

[a] Recovery through complete method may not have been determined.
[b] Elution with successive 200 ml.-portions of ethyl ether/petroleum ether (v/v).
[c] Pesticides for which method is official AOAC.

Table IV. *Chemical names of pesticides mentioned in text*

Pesticide	Chemical name
aldrin	1,2,3,4,10,10-hexachloro-1,4,4a,5,8,8a-hexahydro-1,4-*endo,exo*-5,8-dimethanonaphthalene
azinphosmethyl	*O,O*-dimethyl *S*-(4-oxobenzotriazino-3-methyl) phosphorodithioate
benefin	*N*-butyl-*N*-ethyl-α,α,α,-trifluoro-2,6-dinitro-p-toluidine
BHC	a mixture of isomers of 1,2,3,4,5,6-hexachlorocyclohexane
bromophos-ethyl	*O,O*-diethyl 0-2,5,dichloro-4-bromophenylthionophosphate
bulan	2-nitro-1,1-bis(*p*-chlorophenyl)butane
butoxy ethanol ester 2,4-D	butoxyethyl 2,4-dichlorophenoxyacetate
butoxy ethanol ester 2,4,5-T	butoxyethyl 2,4,5-trichlorophenoxyacetate
n-butyl ester 2,4-D	*n*-butyl 2,4-dichlorophenoxyacetate
n-butyl ester 2,4,5-T	*n*-butyl 2,4,5-trichlorophenoxyacetate
captan	*N*-[(trichloromethyl)thio]-4-cyclohexene-1,2-dicarboximide
carbophenothion	*O,O*-diethyl *S*-(*p*-chlorophenylthiomethyl phosphorodithioate
chlorbenside	*p*-chlorobenzyl *p*-chlorophenyl sulfide
chlordane (tech.)	*cis* and *trans* isomers of 1,2,4,5,6,7,8,8-octachloro-2,3,3a,4,7,7a-hexahydro-4,7-methanoindene plus approximately 50% related chlorinated compounds.
chlordecone	1,2,3,4,5,6,7,8,9,10-decachloro [5.2.1.$0^{2,6}$.$0^{3,9}$.$0^{5,8}$]-decano-4-one
chlorobenzilate	ethyl 4,4′-dichlorobenzilate
chloropropylate	isopropyl 4,4′-dichlorobenzilate
Chlorothion®	*O,O*-dimethyl *O*-(3-chloro-4-nitrophenyl) phosphorothioate
CIPC	isopropyl 3-chlorophenylcarbamate
coumaphos	*O,O*-diethyl *O*-3-chloro-4-methyl-2-oxo-2H-1-benzopyran-7-yl phosphorothioate
Dacthal®	dimethyl tetrachloroterephthalate
o,p′-DDE	1,1-dichloro-2-(*o*-chlorophenyl)-2-(*p*-chlorophenyl)ethene
p,p′-DDE	1,1-dichloro-2,2-bis(*p*-chlorophenyl)ethene
o,p′-DDT	1,1,1-trichloro-2-(*o*-chlorophenyl)-2-(*p*-chlorophenyl)ethane
p,p′-DDT	1,1,1-trichloro-2,2-bis(*p*-chlorophenyl)ethane
DEF®	*S,S,S*-tributyl phosphorotrithioate
dialifor	*O,O*-diethyl-*S*-(2-chloro-1-phthalimidoethyl) phosphorodithioate
diazinon	*O,O*-diethyl *O*-(2-isopropyl-4-methyl-6-pyrimidinyl) phosphorothioate
dichlobenil	2,6-dichlorobenzonitrile
dichloran	2,6-dichloro-4-nitroaniline
dicofol	4,4′-dichloro-α-trichloromethylbenzhydrol
dieldrin	1,2,3,4,10,10-hexachloro-6,7-epoxy-1,4,4a,5,6,7,8a-octahydro-1,4-*endo,exo*-5,8-dimethanonaphthalene

Table IV. (Continued)

Pesticide	Chemical name
Dilan® (mixture of bulan and prolan)	mixture of 2-nitro-1,1-bis(*p*-chlorophenyl)butane and 2-nitro-1,1-bis(*p*-chlorophenyl)propane
diuron	3-(3,4-dichlorophenyl)-1,1-dimethylurea
DMPA	*O*-(2,4-dichlorophenyl) *O*-methyl-isopropylphosphoramidothioate
Dursban®	*O,O*-diethyl 0-3,5,6-trichloro-2-pyridyl phosphorothioate
Dyfonate®	*O*-ethyl-*S*-phenyl-ethylphosphonodithioate
Dyrene®	2,4-dichloro-6-(*o*-chloroanilino)-*s*-triazine
endosulfan I	6,7,8,9,10,10-hexachloro-1,5,5a,6,9,9a-hexahydro 6,9-methano-2,4,3-benzodioxathiepin-3-oxide
endosulfan II	isomer of endosulfan I (about the 3-oxide position)
endrin	1,2,3,4,10,10-hexachloro-6,7-epoxy-1,4,4a,5,6,7,8, 8a-octahydro-1,4-*endo,endo*-5,8-dimethano-naphthalene
endrin alcohol	1,9,10,10,11,12-hexachlorohexacyclo-$[7.2.1.0^{2,6}.0^{9,8}.0^{7,11}.0^{3,12}]$dodecan-3-one
endrin aldehyde	4,5,6,7,8,8-hexachlorohexahydro-4,7-methano-3,-5,6-methenoidan-1-carboxyaldehyde
endrin ketone (Delta keto 153)	1,9,10,10,11,12-hexachloropentacyclo-$[7.2.1.0^{2,6}.0^{4,8}.0^{7,11}]$ dodecan-3-one
EPN	*O*-ethyl *O*-*p*-nitrophenyl phenylphosphonothioate
ethion	*O,O,O′,O′*-tetraethyl *S*, *S′*-methylene bisphos-phorodithioate
ethyl hexyl ester 2,4-D	2-ethylhexyl-2,4-dichlorophenoxyacetate
fenchlorphos	*O,O*-dimethyl *O*-(2,4,5-trichlorophenyl) phos-phorothioate
fenitrothion	*O,O*-dimethyl *O*-(4-nitro-*m*-tolyl) phosphorothioate
folpet	*N*-[(trichloromethyl)thio]phthalimide
genite	2,4-dichlorophenyl benzene sulfonate
heptachlor	1,4,5,6,7,8,8-heptachloro-3a,4,7,7a-tetrahydro-4,7-*endo*-methanoindene
heptachlor epoxide	1,4,5,6,7,8,8-heptachloro-2,3-epoxy-2,3,3a,4,7,7a-hexahydro-4,7-*endo*-methanoindene
hexachlorobenzene	hexachlorobenzene
hexachlorophene	bis(3,5,6-trichloro-2-hydroxyphenyl)methane
isobenzan	1,3,4,5,6,7,8,8-octachloro-3a,4,7,7a-tetrahydro-4, 7-methanophthalan
isobutyl ester 2,4-D	isobutyl 2,4 dichlorophenoxyacetate
isodrin	1,2,3,4,10,10-hexachloro-1,4,4a,5,6,7,8,8a-octahydro-1,4-*endo,endo*-5,8-dimethano-naphthalene
iso-octyl ester 2,4-D	2-ethylhexyl 2,4-dichlorophenoxyacetate
iso-octyl ester 2,4,5-T	2-ethylhexyl 2,4,5-trichlorophenoxyacetate
isopropyl ester 2,4-D	isopropyl 2,4-dichlorophenoxyacetate
isopropyl ester 2,4,5-T	isopropyl 2,4,5-trichlorophenoxyacetate
lindane	1,2,3,4,5,6-hexachlorocyclohexane, gamma isomer
malathion	*O,O*-dimethyl *S*-[1,2-di(ethoxycarbonyl)ethyl] phosphorodithioate
Merphos®	*S,S,S*-tributyl phosphorotrithioite

Table IV. (Continued)

Pesticide	Chemical name
methoxychlor	1,1,1-trichloro-2,2-bis(*p*-methoxyphenyl)ethane
methyl parathion	*O,O*-dimethyl *O-p*-nitrophenyl phosphorothioate
Methyl Trithion®	*O,O*-dimethyl *S*-(*p*-chlorophenylthio) methyl phosphorodithioate
mirex	dodecachloro-octahydro-1,3,4-metheno-2H-cyclobuta(cd)pentalene
monuron	3-(*p*-chlorophenyl)-1,1-dimethylurea
neburon	1-butyl-3-(3,4-dichlorophenyl)-1-methylurea
Nemacide®	*O,O*-diethyl *O*-2,4-dichlorophenylphosphorothioate
octachlor epoxide	1,2,4,5,6,7,8,8-octachloro-2,3-epoxy-2,3,3a,4,7,7a-hexahydro-4,7-methanoindene
ovex	*p*-chlorophenyl *p*-chlorobenzenesulfonate
parathion	*O,O*-diethyl *O-p*-nitrophenyl phosphorothioate
pentachlorophenol	pentachlorophenol
Perthane®	1,1-dichloro-2,2-bis(*p*-ethylphenyl)ethane
Perthane® olefin	1-chloro-2,2-bis(*p*-ethylphenyl)ethene
phencapton	*O,O*-dimethyl *S*-(2,5-dichlorophenylthiomethyl) phosphorodithioate
phorate	*O,O*-diethyl *S*-(ethylthiomethyl)phosphorodithioate
phosalone	*O,O*-diethyl *S*-(6-chloro-2-oxo-benzoxazolin-3-yl)-methyl phosphorodithioate
photodieldrin	1,1,2,3,3a,7a-hexachloro-5,6-epoxy-decahydro-2,4,7-metheno-1H-cyclopenta (a)-pentalene
polychlorinated biphenyls	biphenyl polychlorinates (approximately 20–60% chlorine, various members)
prolan	2-nitro-1,1-bis(*p*-chlorophenyl)propane
quintozene	pentachloronitrobenzene
SD 7438®	toluene-*α,α*-diethyl bis (*O,O*-dimethyl phosphorodithioate)
Strobane®	terpene polychlorinates (contains 65% chlorine)
sulfotepp	*O,O,O,O*-tetraethyl dithiopyrophosphate
Sulphenone®	*p*-chlorophenyl phenyl sulfone
o,p'-TDE	1,1-dichloro-2-*o*-chlorophenyl)-2-(*p*-chlorophenyl)ethane
p,p'-TDE	1,1-dichloro-2,2-bis(*p*-chlorophenyl)ethane
p,p'-TDE olefin	1-chloro-2,2-bis(*p*-chlorophenyl)ethene
tecnazene	1,2,4,5-tetrachloro-3-nitrobenzene
tetradifon	*p*-chlorophenyl 2,4,5-trichlorophenyl sulfone
tetraiodoethylene	tetraiodoethylene
tetrasul	*p*-chlorophenyl 2,4,5-trichlorophenyl sulfide
toxaphene	chlorinated camphene (contains 67–69% chlorine)
trifluralin	*α,α,α*-trifluoro-2,6-dinitro-*N,N*-dipropyl-*p*-toluidine
Vegedex®	2-chloroallyl *N,N*-diethyldithiocarbamate

Acknowledgment

The author recognizes that contributions of various individuals may have been overlooked in this attempt to review the development of FDA's multiresidue methodology. This is particularly true of FDA district laboratory personnel who, through everyday use of analytical procedures, may have made significant undocumented suggestions, observations, or experimentation leading to modifications or improvements. The value of the residue analyst in influencing the development of analytical methods must be emphasized.

I wish to acknowledge the tremendous task and accomplishment of Miss Helen Barry and Mrs. Jean Gaul in supervising the January 1, 1968, revision of the FDA Pesticide Analytical Manual. The PAM is a significant step toward standardization of good pesticide residue analytical methodology. It brings into the analytical laboratory the practical results of the investigations covered in this review.

I also wish to thank Dr. Walter Benson and Mr. Howard Jones for assistance with chemical nomenclature presented in Table IV.

Summary

As discussed in the introduction, the most current and complete description of the Food and Drug Administration analytical method for multiple pesticide residues in fatty and nonfatty types of samples, and of supporting data, is given in the FDA "Pesticide Analytical Manual," Volume I, January 1968, July 1969, and June 1970 revisions. Revisions of the PAM to include new developments will be made once or twice a year. "Official Methods of Analysis" of the AOAC [11th Edition (1970)] contains a detailed description of extraction, cleanup, GLC, and TLC procedures recognized as official AOAC methodology. Improvements and additions to this methodology will be covered in reports of the General Referee for Chlorinated Pesticides, the General Referee for Collaborative Studies of Multi-Residue Methods, and the AOAC Subcommittee E to future meetings of the AOAC.

The material presented, a review of the development of FDA's methodology for multiple residues of organochlorine pesticides, and references to reports containing supporting or validating data, bring together the results of the vast research effort which has gone into the development of this methodology. It shows the extent of refinement of the methodology and should serve as a source for information about the method: its scope, capabilities, and limitations.

No other residue analytical method has been developed to this extent. There will be a continual effort to refine and extend the FDA method, with changes appearing in the FDA "Pesticide Analytical Manual," Volume I.

Methodology included in PAM I, Chapter 2, Sections 211, 212, 231, and 232, Chapter 3 and Chapter 4, Sections 410, 411, and 412 is, in the author's opinion, the best available for the determination of multiple organochlorine (and certain organophosphorus) pesticide residues in foods and feeds.

Résumé*

Développement de la méthode de la Food and Drug Administration pour l'analyse polyvalente des résidus de pesticides organochlorés dans les aliments et les fourrages

Comme il a été indiqué dans l'introduction, la description la plus connue et la plus complète de la méthode de la Food and Drug Administration pour l'analyse polyvalente des résidus de pesticides dans les échantillons de matières alimentaires à forte ou faible teneur en matières grasses, est donnée dans le Manuel analytique des pesticides (Pesticides Analytical Manual) de la F.D.A., volume I de janvier 1968 et les révisions de juillet 1969.

Pour introduire de nouvelles techniques, des révisions de ce manuel seront faites une ou deux fois par an. Les "Changements dans les méthodes officielles" de l'A.O.A.C (Journ. of A.O.A.C., 1968, **51**, 473–81) renferment une description détaillée de l'extraction, de la purification, des techniques par chromatographie gazeuse et en couche mince reconnues commes officielles par l'A.O.A.C. Des perfectionnements récents et des compléments à ces méthodes seront introduits dans la 11 ème édition 1970 des méthodes officielles de l'A.O.A.C.

Ces changements seront décrits dans les compte rendus du Rapporteur général des pesticides organochlorés, du Rapporteur général des études en collaboration de la méthode d'analyse polyvalente des résidus et le sous comité E de l'A.O.A.C. à la réunion de 1969.

Cette mise au point sur les développements des techniques de la F.D.A. pour l'analyse polyvalente des résidus organochlorés et les références aux travaux sur ce sujet qui y sont rapportés rassemblent les résultats des larges efforts de recherche suscités par l'elaboration de ces techniques. Elle montre la valeur de la méthode et précise son domaine, ses possibilités et ses limites.

Aucune autre méthode d'analyse des résidus n'a été aussi étudiée. Un effort continu d'amélioration et d'extension sera poursuivi, avec des changements qui interviendront dans le Manuel analytique des pesticides, volume I.

Les techniques décrites dans le P. A. M. I, Chapitre 2, sections 211–212–231 et 232, Chapitre 3 et Chapitre 4, sections 410–411 et 412 sont, de l'avis de l'auteur, les meilleures dont on dispose actuellement

* Traduit par R. MESTRES.

pour la détermination polyvalente des résidus de pesticides organochlorés (et de certains organophosphorés) dans les aliments et les fourrages.

Zusammenfassung*

Entwicklung der Methodik der Food and Drug Administration zur "multiple residues"—Analyse von Organochlor-Pestiziden in Lebens-und Futtermitteln

Wie in der Einleitung besprochen, enthält das "Pesticide Analytical Manual" der Food and Drug Administration (FDA), Vol. I, Januar 1968, Stand Juli 1969, die neueste und vollständigste Beschreibung der "multiple residues"—Analysenverfahren der FDA für fetthaltige und fettfreie Proben. Das "Pesticide Analytical Manual" wird ein- bis zweimal jährlich unter Einbeziehung neuer Entwicklungen überarbeitet. Die AOAC–Schrift "Changes in Official Methods" (J. Assoc. Official Anal. Chemists *51*, 473–481 (1968)) enthält eine ausführliche Beschreibung der als offizielle AOAC-Methodik anerkannten Extraktions–, Reinigungs–, gas– und dünnschichtchromatographischen Verfahren. Neuere Verbesserungen und Ergänzungen dieser Methodik sind in der bevorstehenden 11. Auflage (1970) der "Official Methods of Analysis" der AOAC enthalten. Diese Änderungen werden in Berichten des Referenten für Chlorierte Pestizide, des Referenten für Gemeinschafts–Studien über "multiple residues"—Verfahren und des Unterausschusses E der AOAC an die AOAC–Jahrestagung 1969 behandelt.

Das vorgelegte Material gibt auf Grund von Sachberichten einen Überblick über die Entwicklung der FDA–Analysen-methodik und die umfangreiche Forschung, die zur Entwicklung dieser Methoden erforderlich war. Sie zeigt, wie die Methoden verfeinert wurden und sollte als Quelle für die Information über Umfang, Möglichkeiten und Grenzen der "multiple residues"—Methodik dienen.

Kein anderes Verfahren der Rückstandsanalytik wurde so weitgehend entwickelt wie die FDA "multiple residues"—Methodik. Es ist beabsichtigt diese laufend zu verfeinern und zu erweitern. Änderungen erscheinen in dem "Pesticide Analytical Manual," Vol. 1, der FDA.

Nach Ansicht des Verfassers sind die in Kapitel 2, Abschnitt 211, 212, 231 und 232 und die in den Kapiteln 3 und 4, Abschnitt 410, 411 und 412 des "Pesticide Analytical Manual," Vol. 1, beschriebenen Methoden die zur Zeit besten Verfahren zur Bestimmung multipler Organochlor– (und bestimmter Organophosphor–) Pestizidrückstände in Lebens– und Futtermitteln.

---

* Übersetzt von H. MAIER-BODE.

References

Armour, J. A., and J. A. Burke: Polychlorinated biphenyls as potential interference in pesticide residue analysis. Private communication, *U. S. Food and Drug Administration* (July 1969).

——— ——— A method for separating polychlorinated biphenyls from DDT and its analogs. J. Assoc. Official Anal. Chemists **53,** 761 (1970).

Banks, K., and D. Bills: Gas chromatographic identification of chlorinated insecticides based on their UV degradation. J. Chromatog. **33,** 450 (1968).

Beroza, M., and M. C. Bowman: Extraction of insecticides for cleanup and identification. J. Assoc. Official Agr. Chemists **48,** 358 (1965).

Bertuzzi, P. F., L. Kamps, C. I. Miles, and J. A. Burke: Extraction of chlorinated pesticide residues from nonfatty samples of low moisture content. J. Assoc. Official Anal. Chemists **50,** 623 (1967).

Boggs, H. M., and G. Q. Lipscomb: Unpublished data, *U. S. Food and Drug Administration* (Dec. 1961).

Bowman, M. C., and M. Beroza: Extraction p-values of pesticides and related compounds in six binary solvent systems. J. Assoc. Official Agr. Chemists **48,** 943 (1965).

Burke, J. A.: Report on chlorinated insecticides and miticides. J. Assoc. Official Anal. Chemists **50,** 575 (1967).

——— Report on chlorinated pesticides. J. Assoc. Official Anal. Chemists **51,** 311 (1968).

——— The analytical characteristics of two dieldrin photoconversion products. Bull. Environ. Contamination Toxicol. **4,** 152 (1969).

———, and L. Giuffrida: Investigations of electron capture gas chromatography for the analysis of multiple chlorinated pesticide residues in vegetables. J. Assoc. Official Agr. Chemists **47,** 326 (1964).

———, and W. Holswade: Gas chromatography with microcoulometric detection for pesticide residue analysis. J. Assoc. Official Agr. Chemists **47,** 845 (1964).

——— ——— A gas chromatographic column for pesticide residue analysis: Retention times and response data. J. Assoc. Official Anal. Chemists **49,** 374 (1966).

———, and L. Johnson: (I) Cleanup in the use of microcoulometric gas chromatographic apparatus for pesticide residue analysis. (II) Microcoulometric gas chromatography of pesticides. 140th Nat. Meeting Amer. Chem. Soc., Chicago Sept. 3–8 (1961).

——— ——— Investigations in the use of the microcoulometric gas chromatograph for pesticide residue analysis. J. Assoc. Official Agr. Chemists **45,** 348 (1962).

———, and B. Malone: Effects of calcination temperature and time on the retentive properties of Florisil used for pesticide residue analysis. J. Assoc. Official Anal. Chemists **49,** 1003 (1966).

———, and P. A. Mills: Microcoulometric gas chromatographic determination of Thiodan and Tedion in green vegetables. J. Assoc. Official Agr. Chemists **46,** 177 (1963).

———, and M. L. Porter: A study of the effectiveness of some extraction procedures for pesticide residues in vegetables. J. Assoc. Official Anal. Chemists **49,** 1157 (1966).

——— ——— Note on the effect of sample moisture content on extraction of TDE from kale. J. Assoc. Official Anal. Chemists **50,** 1260 (1967).

——— ———, and S. J. V. Young: Extraction of foliar and root adsorbed pesticide residues from fresh and dehydrated nonfatty samples by the Mills, Onley, Gaither acetonitrile procedure. Submitted to J. Assoc. Official Anal. Chemists.

Carr, R. L.: Collaborative study of the Mills method for multiple chlorinated pesticide residues in butterfat. J. Assoc. Official Anal. Chemists **53,** 152 (1970).

——— Collaborative study of a method for analysis of fish for multiple residues of chlorinated pesticides. Submitted to J. Assoc. Official Anal. Chemists.

CHAU, A. S. Y.: Derivative formation for confirmation of endosulfan by gas chromatography. J. Assoc. Official Anal. Chemists **52**, 1240 (1969).

———, and W. P. COCHRANE: Cyclodiene Chemistry. I. Derivative formation for the identification of heptachlor, heptachlor epoxide, *cis*-chlordane, *trans*-chlordane, and aldrin pesticide residues by gas chromatography. J. Assoc. Official Anal. Chemists **52**, 1092 (1969 a).

——— ——— III. Derivative formation for the identification of heptachlor, heptachlor epoxide, *cis*-chlordane, *trans*-chlordane, dieldrin, and endrin pesticide residues by gas chromatography. J. Assoc. Official Anal. Chemists **52**, 1220 (1969 b).

COCHRANE, W. P., and A. S. Y. CHAU: Note on gas chromatographic identification of heptachlor pesticide residues by derivative formation. J. Assoc. Official Anal. Chemists **51**, 1267 (1968).

COULSON, D. M., L. A. CAVANAGH, J. E. DEVRIES, and B. WALTHER: Microcoulometric gas chromatography of pesticides. J. Agr. Food Chem. **8**, 399 (1960).

DAVIDSON, A.: Collaborative study of a rapid method for multiple chlorinated pesticide residues in small fruits. J. Assoc. Official Anal. Chemists **49**, 468 (1966).

DUFFY, J. R., and N. WONG: Residues of organochlorine insecticides and their metabolites in soils in the Atlantic Provinces of Canada. J. Agr. Food Chem. **15**, 457 (1967).

FITZ-SIMMONS, O.: U. S. Patent No. 2,393,625 (Jan. 29, 1946).

Floridin Company: Tech. Bull. "Florisil" (1968).

FURMAN, W. B., and N. V. FEHRINGER: Study of modifications of the Mills procedure for chlorinated pesticides in dairy products. J. Assoc. Official Anal. Chemists **50**, 903 (1967).

GARDNER, A.: The separation and detection of ronnel, ethion, and carbophenothion on Silica Gel G thin layer chromatographic plates. Private communication *U. S. Food and Drug Administration* (Aug. 1967).

——— Confirmation of organophosphorus pesticide residues at nanogram levels by two-dimensional thin layer chromatography. Submitted to J. Assoc. Official Anal. Chemists.

GAUL, J.: Collaborative study of a method for multiple chlorinated pesticide residues in leafy and cole-type vegetables. J. Assoc. Official Anal. Chemists **49**, 463 (1966).

——— Investigation of Florisil and quality of ethyl and petroleum ethers as a cause of low recoveries of certain pesticides. Private communication *U. S. Food and Drug Administration* (May 1969 a).

——— Low results encountered in the acetonitrile dilution step in the "Rapid method for chlorinated pesticide residues in nonfatty foods." Private communication *U. S. Food and Drug Administration* (May 1969 b).

GIUFFRIDA, L.: A flame ionization detector highly selective and sensitive to phosphorus—A sodium thermionic detector. J. Assoc. Official Agr. Chemists **47**, 293 (1964).

———, and N. F. IVES: Improvements in the gas chromatography column used for pesticide analysis. Private communication *U. S. Food and Drug Administration* (Sept. 1965).

——— ——— Microcoulometric gas chromatography: Improvements in instrumentation. J. Assoc. Official Anal. Chemists **52**, 541 (1969).

GREENSPAN, N. E., S. M. HART, and R. J. MARTIN: Gas chromatography with electron capture detection. Private communication *U. S. Food and Drug Administration* (Nov. 1962).

GUNTHER, F. A., and R. C. BLINN: Analysis of insecticides and acaricides, p. 46. New York: Interscience (1955).

HAMMENCE, J. H., P. S. HALL, and D. J. CAVERLY: The identification and determination of chlorinated pesticide residues. Analyst **90**, 649 (1965).

HORWITZ, W.: Official Methods of Analysis of the Association of Official Analytical Chemists (AOAC) (1970).

——— Changes in Official Methods of Analysis Made at the Eighty-first Annual Meeting Oct. 9–12, 1967. J. Assoc. Official Anal. Chemists **51**, 470 (1968).

IVES, N. F., and L. GIUFFRIDA: Unpublished data (Feb. 1968).

JOHNSON, L.: Separation of dieldrin and endrin from other chlorinated pesticide residues. J. Assoc. Official Agr. Chemists **45**, 363 (1962).

——— Collaborative study of a method for multiple chlorinated pesticide residues in fatty acids. J. Assoc. Official Agr. Chemists **48**, 668 (1965).

JOHNSON, R. H., and H. R. THROM: Clean-up of fat samples. Private communication. *U. S. Food and Drug Administration* (Dec. 1964).

JONES, L. R., and J. A. RIDDICK: Separation of organic insecticides from plant and animal tissues. Anal. Chem. **24**, 569 (1952).

KLEIN, A. K.: Report on extraction procedures for chloroorganic pesticides. J. Assoc. Official Agr. Chemists **41**, 551 (1958).

———, E. P. LAUG, and J. D. SHEEHAN, JR.: Report on extraction procedures for chloroorganic insecticides. J. Assoc. Official Agr. Chemists **42**, 539 (1959).

———, and J. O. WATTS: Separation and measurement of Perthane, DDD (TDE) and DDT in leafy vegetables by electron capture gas chromatography. J. Assoc. Official Agr. Chemists **47**, 311 (1964).

KOVACS, M. F.: Thin layer chromatography for chlorinated pesticide residue analysis. J. Assoc. Official Agr. Chemists **46**, 884 (1963).

——— Thin layer chromatography for organothiophosphate pesticide residue determination. J. Assoc. Official Agr. Chemists **47**, 1097 (1964).

KRAUSE, R.: Collaborative study of a method for multiple chlorinated pesticide residues in nonfatty vegetables. J. Assoc. Official Anal. Chemists **49**, 460 (1966).

LASKI, R. L.: GLC determination of BHC. Private communication *U. S. Food and Drug Administration* (Aug. 1969).

LAWRENCE, J. H.: Relative retention times and response characteristics of chlorinated and organophosphorus pesticides and related compounds—Isothermal electron capture GLC. Private communication *U. S. Food and Drug Administration* (June 1966).

———, and J. A. BURKE: Comparison of ten methods for the analysis of milk for residues of chlorinated pesticides. J. Assoc. Official Anal. Chemists **52**, 817 (1969).

———, and J. ROACH: Relative retention times and response characteristics of chlorinated and organophosphorus pesticides and related compounds—Isothermal electron capture GLC. Private communication *U. S. Food and Drug Administration* (Mar. 1967).

LIPSCOMB, G. Q.: Electron capture gas chromatography. Private communication *U. S. Food and Drug Administration* (Feb. 1963).

LOVELOCK, J. E., and S. W. LIPSKY: Electron affinity spectroscopy—A new method for the identification of functional groups in chemical compounds separated by gas chromatography. J. Amer. Chem. Soc. **82**, 431 (1960).

MALONE, B.: Preparation of ethyl ether for use in Florisil column chromatography eluants. Private communication. *U. S. Food and Drug Administration* (Nov. 1966).

MILLER, G. A.: Pesticide retention data for a 3% OV-17 on Gas Chrom Q GLC column. Private communication. *U. S. Food and Drug Administration* (June 1970).

———, and C. E. WELLS: Alkaline pre-column for use in gas chromatographic pesticide residue analysis. J. Assoc. Official Anal. Chemists **52**, 548 (1969).

MILLS, P. A.: Detection and semiquantitative estimation of chlorinated organic pesticide residues in foods by paper chromatography. J. Assoc. Official Agr. Chemists **42**, 734 (1959).

——— Collaborative study of certain chlorinated organic pesticides in dairy products. J. Assoc. Official Agr. Chemists **44**, 171 (1961).

——— Variation of Florisil activity: Simple method for measuring adsorbent capacity and its use in standardizing Florisil columns. J. Assoc. Official Anal. Chemists **51**, 29 (1968).

———, and J. A. BURKE: Unpublished data *U. S. Food and Drug Administration* (Oct. 1964).

———, J. H. ONLEY, and R. A. GAITHER: Rapid method for chlorinated pesticide residues in nonfatty foods. J. Assoc. Official Agr. Chemists **46**, 186 (1963).

MITCHELL, L. C.: Note on the detection of chlorinated organic pesticides in the paper chromatogram. J. Assoc. Official Agr. Chemists **39**, 891 (1956).

——— Ascending paper chromatography—A way to do it. J. Assoc. Official Agr. Chemists **40**, 999 (1957).

MODDES, R. E. J., and J. W. COOK: The extraction and identification of parathion and diazinon from lettuce. J. Assoc. Official Agr. Chemists **42**, 208 (1959).

MOYA, H.: Clean-up of fatty food samples for pesticide residue analysis. Private communication. *U. S. Food and Drug Administration* (Mar. 1967).

MUMMA, R O., W. B. WHEELER, D. E. H. FREAR, and R. H. HAMILTON: Dieldrin: Extraction of accumulations by root uptake. Science **152**, 530 (1966).

NELSON, R. C.: Screening procedures for organothiophosphate pesticide residues on fruits and vegetables by microcoulometric gas chromatography. J. Assoc. Official Agr. Chemists **47**, 289 (1964).

ONLEY, J. H., and P. A. MILLS: Detection and estimation of chlorinated pesticides in eggs. J. Assoc. Official Agr. Chemists **45**, 983 (1962).

———, and G. YIP. Analysis of a single crop extract for substituted urea herbicides and metabolites, chlorinated insecticides and Amitrole. J. Assoc. Official Anal. Chemists **52**, 526 (1969).

OSADCHUK, A. A., and E. B. WANLESS: Identification and quantitative estimation of aldrin residues in the presence of interfering materials in samples of plant and animal origin. J. Assoc. Official Anal. Chemists **51**, 1264 (1968).

PARDUE, J.: Recovery of organophosphorus pesticides by the AOAC multiresidue method. Submitted to J. Assoc. Official Anal. Chemists.

POMERANTZ, I. H., L. J. MILLER, and C. KAVA: Extraction, cleanup, and gas liquid chromatographic method for captan, folpet and Difolatan in crops. J. Assoc. Official Anal. Chemists **53**, 154 (1970).

PORTER, M. L., and J. A. BURKE: A preliminary study of the extraction of parathion and diazinon by the Mills, Onley, Gaither acetonitrile procedure. J. Assoc. Official Anal. Chemists **51**, 63 (1968).

——— ——— Modification of the AOAC method for multiple pesticide residues in nonfatty foods for samples of high sugar content. J. Assoc. Official Anal. Chemists **52**, 1280 (1969).

——— ———, and P. F. BERTUZZI: Acetonitrile dilution: A possible source of error in the "Rapid method for chlorinated pesticide residues in nonfatty foods." J. Assoc. Official Anal. Chemists **50**, 644 (1967).

———, R. GAJAN, and J. A. BURKE: Acetonitrile extraction and determination of carbaryl in fruits and vegetables. J. Assoc. Official Anal. Chemists **52**, 177 (1969).

———, S. J. V. YOUNG, and J. A. BURKE: Unpublished data *U. S. Food and Drug Administration* (June 1969).

——— ——— ——— Analysis of fish, animal, and poultry tissue for chlorinated pesticide residues. Submitted to J. Assoc. Official Anal. Chemists.

SANS, W. W.: Multiple insecticide residues determination using column chromatography, chemical conversion, and gas liquid chromatography. J. Agr. Food Chem. **15**, 192 (1967).

SAWYER, A. D.: Study of extraction procedures for chlorinated organic pesticides in fresh eggs. J. Assoc. Official Anal. Chemists **49**, 643 (1966).

SHUMAN, H.: Unpublished data *U. S. Food and Drug Administration* (May 1963).
STORHERR, R. W., and J. A. BURKE: Determination of 3-amino-1,2,4-triazole in crops. J. Assoc. Official Agr. Chemists **44**, 196 (1961).
———, P. OTT, and R. R. WATTS: A general method for organophosphorus residues in non-fatty foods. Submitted to J. Assoc. Official Anal. Chemists.
THOMAS, E., J. A. BURKE, and J. H. LAWRENCE: Thin layer chromatography: Relative migration data (R_{TDE}) of chlorinated pesticides. J. Chromatog. **35**, 119 (1968).
WATTS, J. O., and A. K. KLEIN: Determination of chlorinated pesticide residues by electron capture gas chromatography. J. Assoc. Official Agr. Chemists **45**, 102 (1962).
WATTS, R. R., R. W. STORHERR, J. R. PARDUE, and T. OSGOOD: Charcoal column cleanup method for many organophosphorus pesticide residues in crop extracts. J. Assoc. Official Anal. Chemists **52**, 522 (1969).
WELLS, C. E.: A rapid gas chromatographic screening procedure for pesticide residues on fruits and vegetables. Private communication *U. S. Food and Drug Administration* (Aug. 1962).
——— Validation study of a method for pesticide residues in foods and animal feeds. J. Assoc. Official Anal. Chemists **50**, 1205 (1967).
WESSEL, J. R.: Collaborative study of a method for multiple organophosphorus pesticide residues in nonfatty foods. J. Assoc. Official Anal. Chemists **50**, 430 (1967).
——— Collaborative study of three gas chromatographic dual detection systems for analysis of multiple chlorinated and organophosphorus pesticides. J. Assoc. Official Anal. Chemists **51**, 666 (1968).
——— Comparison of four extraction and cleanup procedures for chlorinated pesticide residues in eggs. J. Assoc. Official Anal. Chemists **52**, 172 (1969).
WIENCKE, W. W., and J. A. BURKE: Losses of organophosphorus pesticides due to oxidation during sample concentration. Private communication *U. S. Food and Drug Administration* (Feb. 1969 a).
——— Derivatization of dieldrin and endrin for confirmation of residue identity. J. Assoc. Official Anal. Chemists **52**, 1277 (1969 b).
WILDERMAN, M., and H. SHUMAN: Extraction procedures for DDT residues in firm fruits and in hay: Colorimetric method. J. Assoc. Official Anal. Chemists **51**, 892 (1968).

Manuscript received 8 January 1970; accepted 28 February 1970.

The metabolism of contact insecticides in stored grains. II. 1966–1969*

By

D. G. Rowlands**

Contents

I. Introduction

In many developing countries the background of increasing population has not been matched by increased food production or distribution from over-producers. In consequence the general tendency to surplus grain production, characteristic of world markets in the '50s, had become one of near balance between supply and demand by the mid '60s (UN/FAO 1966). However, the heavy 1969 crops of cereal grains (UN/FAO 1969) should offset this depletion somewhat and will probably lead to bulk storage, requiring long-term protective treatment with insecticides.

* Part I, see Rowlands, D. G.: The metabolism of contact insecticides in stored grains. Residue Reviews 17, 105 (1967). Present literature survey up to December 31, 1969.

** Pest Infestation Control Laboratory, Slough, England.

Many countries now require their imports to have a high standard of cleanliness and to be free from infestation. This has resulted in increased use of fumigants and contact insecticides. Current regulations in Australia and the Argentine require that all exported wheat be treated with malathion, for example.

With the increased use of grain protectants one can expect increased problems of insecticide resistance in storage pests which are not only widespread but are moved from country to country and from commodity to commodity through trade channels (DYTE and BLACKMAN 1970). There are already instances of resistance to carbamates, some fumigants, pyrethrum, DDT, lindane, and malathion among stored-product insects (PARKIN 1965, SPEIRS *et al.* 1967, CHAMP and CAMPBELL-BROWN 1970).

The long-term answer to resistance and to the toxicological problem of insecticide residues in raw and processed cereals is likely to be the control of insects by physical and genetical means or, possibly, by the use of attractants, repellents, sterilants, or growth inhibitors. The stored grain environment might favour the use of attractants and there may be a case for more studies on the pheromones of stored-product insects.

The toxicological significance of terminal residues (pesticides and their degradation products) on stored foodstuffs is being studied at international levels and attention is being given to the establishment of residue tolerances for both raw and processed cereal grains (UN/FAO/WHO 1969). Evaluation of existing information has shown the urgent need for more intensive work on the nature of terminal residues and their fates in cereal products following milling, doughmaking, and baking. HURTIG (1966) has described the structure and activities of the bodies concerned in establishing workable tolerances in cereals for international use: the IUPAC Pesticides Section, the joint FOA/WHO Expert Committee on Pesticide Residues and the FAO/WHO Codex Alimentarius Committee on Pesticide Residues.

The previous review (ROWLANDS 1967) provided background information on the metabolism of insecticides by stored cereal grains and, more particularly, on the environmental and physiological factors concerned in such metabolism. For an understanding of their relevance, and to avoid undesirable repetition here, the present reader is referred to that article. This present report is supplementary, and it is hoped that it reflects current thinking on the subjects associated with pesticide metabolism in stored grain.

II. Insecticides[1] used on stored grain

Malathion, lindane, and synergised pyrethrins are still the most commonly used insecticides for application to storage premises and

[1] The chemical names of insecticidal and some other compounds mentioned in the text are given in Table XIII.

containers, or for admixture with edible grain. Sumithion, DDVP, and bromophos are the most likely alternatives (LEMON 1967, KANE and GREEN 1968, TYLER and GREEN 1968, CHAMP *et al.* 1969). Little further evaluation has been made of other compounds suggested in the previous review (dibrom, Guthion, ronnel, fenthion, imidan, chlorthion, dicapthon, formothion, methyl Trithion, dipterex, and tributphon) apart from diazinon (STRONG *et al.* 1967, CHAMP *et al.* 1969) and few new "candidates" have been suggested (COGBURN 1967, VIKHANSKII 1967). Candidates are likely to be used initially for treating premises only and not for admixture as the toxic hazards are less and "clearance" for such use is therefore more readily obtained; indeed sumithion and bromophos are now permitted for use in storage premises. Some countries, e.g., Yugoslavia, do not allow direct admixture of pesticide with grains for consumption.

Development of resistance in stored-product pests to malathion and lindane, presently known to be spreading (DYTE and BLACKMAN 1970), is most serious because the alternatives are likely to be more expensive, perhaps prohibitively so for the developing countries (malathion and lindane no longer have patent protection). Even more disturbing are reports of a field strain of the flour beetle, *Tribolium castaneum,* encountered in peanuts that is not only resistant to malathion and lindane but apparently to several potential alternatives, (many of them not yet in use) including Gardona, sumithion, diazinon, DDVP, and Sevin (CHAMP and CAMPBELL-BROWN 1970). Such a threat emphasises the need for alternative methods of control, but these have so far hardly been considered, let alone practically assessed, for stored-product pests (JACOBSON 1965, APPLEBAUM and KONIJN 1966, MATHLEIN 1967, WILLIAMS 1969, STANIC and SHULOV 1969). It is in alternative insecticides, cheaper fumigants, and wider use of synergists that the immediate answer will have to be sought.

For effectiveness against some pests a grain protectant should remain on the seed coat of the grain; none of the existing insecticides do so (see section III). A formulation that holds the pesticide on the seed coat would obviously be of value and addition of sodium carboxymethyl cellulose (S-CMC) to emulsion sprays could be a step toward this end (TYLER and ROWLANDS 1968). Consideration of the insect species involved and their feeding habits could also contribute to "specific formulation." Thus the saw-toothed grain beetle, *Oryzaephilus surinamensis,* which feeds almost exclusively on the germ (SURTEES 1964) could well be controlled more efficiently by a compound which accumulates in the germ than by one which passes rapidly to the endosperm. Alternatively, an insecticide that rapidly reaches the endosperm, or that is aided in so doing by the formulation used, might be advantageous to the farmer wishing to kill larvae of the grain weevil, *Sitophilus* spp., before the emergence holes cause downgrading of his grain (GREEN 1969). The insecticide should also be quick acting, so that insects do not have time to disperse or oviposit before receiving a lethal dose. The most urgent requirement in the United Kingdom is for an insecticide that will remain stable and effective on warm, moist grain; these being, of course, the very conditions that accelerate degradation of most pesticides, yet enhance insect, mite, and fungal development (ROWLANDS 1967).

Of the existing classes of insecticides, the available carbamates continue to show little potential against stored grain pests. Use of the persistent organochlorine compounds has happily declined and, although still used extensively in the developing countries, DDT is no longer recommended for direct application to grain. The use of lindane, mainly as a dust formulation, continues much as heretofore.[2] Some new pyrethroids of interest have emerged, the compound NRDC 104 and its (+)-*trans*-isomer (NRDC 107) were synthesised by ELLIOTT *et al.* (1967). Pyrethrin-resistant grain weevils can detoxify these compounds more effectively than the natural pyrethrins but the detoxication is largely blocked by piperonyl butoxide (LLOYD and FIELD 1968 and 1969). Use of these compounds is likely to be restricted by the cost of production and marketing but it is believed that some are commercially available on the Continent and will soon be available in Japan, possibly formulated with pyrethrum or allethrin. Bioallethrin (UCALF 1968), the (+) *trans*-isomer of allethrin, is 1.5 times more toxic to the grain weevil than allethrin, whether synergised or not (LLOYD *et al.* 1970). A Japanese synthetic pyrethroid, Phthalthrin (Neopynamin), is rapidly detoxified by grain weevils and appears to have little prospect for stored-product use unless synergised (LLOYD and FIELD 1969) but has some repellent effects. NRDC 108 seems to have little potential for stored-product insects (LLOYD *et al.* 1970). Two new compounds recently developed in Japan but not yet evaluated are Jasmethrin, a jasmolin analogue, and Prothrin, a propargyl furfuryl chrysanthemate (KATSUDA *et al.* 1970).

Compounds of the organophosphorus group continue to provide the greatest number of candidates. Sumithion (fenitrothion, Folithion, Accothion) continues to show great promise (CHAMP *et al.* 1969, KANE and GREEN 1968, TYLER and GREEN 1968); however, the lack of any exclusive patent controlling its manufacture and marketing may mean that it is not economically viable for very long. Moreover, although the triphenyl phosphate-suppressible, malathion-specific (carboxyesterase type) resistant flour beetles are susceptible to Sumithion (DYTE and ROWLANDS 1968) another nonspecific malathion-resistant strain apparently has some cross tolerance to Sumithion (CHAMP and CAMPBELL-BROWN 1970) and this may be the case with grain storage pests.

---

[2] *Added in proof:* Mention should be made of lindane/malathion mixtures of which at least one (dust) formulation, of 0.4 percent and 2.0 percent respective ingredients, is known to be in commercial production for admixture to grain at recommended levels of two and 10 p.p.m.

DDVP (dichlorvos) has proved excellent for the rapid disinfestation of grain, sometimes at low doses (two p.p.m.), but is a poor residual protectant as it volatizes and breaks down rapidly (GREEN and TYLER 1966). The area of greatest potential for DDVP would seem to be in disinfestation of bagged grain and enclosed grain storage premises. GREEN and WILKIN (1969) have shown that DDVP can be distributed in bagged grain by blowing it through the intergranular spaces so rapidly as to compensate for the high sorptive capacity of the grain (see section III). It has been proposed that DDVP might be used to retreat grain that had been treated previously with an insecticide (say malathion) that had lost effectiveness (STRONG and SBUR 1964). This practice could be hazardous as DDVP may well potentiate the mammalian toxicity of such (malathion) residues as remained [acute oral LD_{50} rat mg./kg.: malathion 1,400–1,900, DDVP 25–30, mixture (4:1) malathion:DDVP 50]. In this connexion, experimental use of a DDVP/malathion mixture for disinfestation of empty railway trucks has been reported (SCHESSER 1967). DDVP undoubtedly prolongs the intact residual life of some organophosphorus insecticides applied to grain and the effect appears to be reciprocal (ROWLANDS 1970 a). CHAMP *et al.* (1969) found antagonism between DDVP-malathion against *Sitophilus oryzae* and *Rizopertha dominica* whereas DDVP-diazinon mixtures were more toxic to *S. oryzae* but not to *R. dominica.*

Bromophos is more stable chemically on damp grain than malathion or sumithion but its effective life is little different. In addition it kills slowly so that treated insects may disperse and oviposit before getting a lethal dose (GREEN *et al.* 1970). The reason for this slow action may be due to difficulties in cellular transport of the toxicant *in vivo* or because an interval for more extensive metabolism is needed. DYTE and ROWLANDS (1968 and 1970), using topical application methods on flour beetles, found that bromoxon (P=O analogue) was no more toxic than bromophos, whereas sumioxon was about 1.5 times more toxic than sumithion and malaoxon was more than three times more toxic than malathion. These differences probably reflect different degrees of ionisation in the "oxons" which affect the amounts transported to the site of lethal action *in vivo*. There may, in addition, be a penetration or "opportunity" barrier that initially restricts the amount of applied bromophos reaching oxidative sites in the insect, since adult flour beetles produced considerably more internal oxons from sumithion and malathion than from bromophos. It is also possible that one or more metabolites beside the oxon contribute to the insecticidal effect of bromophos: compare with the unknown anticholinesterase phosphate found in cattle grubs on cattle treated with the trichlorophenyl analogue, Ronnel (HOPKINS and KNAPP 1963). Despite these shortcomings, bromophos may well be useful if resistance to malathion is encountered in grain pests since neither the malathion-specific (TPP-suppressible) nor, apparently, the non-specific malathion resistant flour beetles, mentioned earlier, show cross-tolerance to bromophos or to its iodo-analogue, iodophos (CHAMP and CAMPBELL-BROWN 1970). There is the additional possibility that halophenyl residues (e.g., from ronnel, bromophos, iodophos) in grain may have some fungistatic or fungicidal effect upon the endemic fungi. Visual examination, after 18 months storage in the laboratory, of damp wheat originally treated at ten p.p.m. with malathion or bromophos, revealed that the same grain treated with bromophos was considerably less mouldy than that treated with malathion (ROWLANDS and CLARKE 1968) but these effects were not observed in the field (GREEN *et al.* 1970).

Iodophos (C-9491, Alphacron, iodfenphos) (HADDOW and MARKS 1969) was included in an assessment by STRONG and SBUR (1968) and is currently one of the insecticides of greatest potential for stored grain. GRADIDGE *et al.* (1970 b) have applied a water-dispersible powder to various building surfaces and found it to show considerable promise against saw-toothed grain beetles. WILKIN *et al.* (1970) have applied iodophos emulsion to barley in three farm trials and a five percent dust in another trial and found it distinctly promising for protective treatment of grain. KANE and AGGARWAL (1970) have used iodophos emulsion on bagged wheat and barley at five and ten p.p.m. and found the treatment still effective after nine weeks.

Cidial [(dime)phenthoate, papthion] appeared to be a useful candidate as, although more active against some species than malathion or lindane, the pure compound had a low mammalian toxicity and showed very considerable persistence on damp wheat (LONGONI and MICHIELI 1967). However, the high mammalian toxicity of the technical product(s) (due possibly to the S-methyl isomer?) means that it is unlikely to be used on grain unless the technical grade material can be rendered less toxic. It is, moreover, the only commercial insecticide to which malathion-specific resistant flour beetles have a cross tolerance (DYTE and BLACKMAN 1970). It may prove of use for treatment of premises (GRADIDGE *et al.* 1970 a) or of bags and sacks containing grain (KANE and AGGARWAL 1970).

Gardona (SD 8447) is another compound that has shown some potential for use on grain (LEMON 1967 a, STRONG and SBUR 1968) because of its reported stability and low mammalian toxicity (WHETSTONE *et al.* 1966). However, several of its (hydrolytic) degradation products are more toxic to mammals than the parent compound, particularly the desmethyl derivative, the trichlorophenacyl chloride (TCC), the trichloroacetophenone (TCA), and the trichlorophenyl ethanol (TCE) (WHETSTONE *et al.* 1966, BEYNON and WRIGHT 1969). It is not very stable on damp wheat, but KANE and AGGARWAL (1970) recently used ten and 20 p.p.m. Gardona on bagged wheat and barley and found these treatments to be effective for more than nine weeks. Although Gardona is effective against the malathion-specific resistant flour beetles the other nonspecific strain referred to seems to have a remarkably high cross tolerance to Gardona (CHAMP and CAMPBELL-BROWN 1970).

Certain diatomaceous earths and silica aerogels have been further assessed (LA HUE and FIFIELD 1967) and powdered neem seed (*Azadirachta indica*) kernels have been admixed with stored wheat as a repellent and protectant (JOTWANI and SIRCAR 1965). MATHLEIN (1967) has evaluated some substituted octyl sulphides and sulfoxides as repellents for admixture with grain and GLYNNE-JONES and SYLVESTER (1966) and WEAVING and SYLVESTER (1967) have reviewed the repellent properties of pyrethrum. APPLEBAUM and KONIJN (1966) have studied the Tribolium protease and amylase inhibitors that occur in wheat grain and which affect the growth and development of flour beetles [see section V a) 2].

III. Factors affecting metabolism and terminal residues

Chemical assay of overall residues of intact pesticide in a given sample taken from a treated grain bulk is not always a reliable indication of chemical persistence or biocidal effectiveness. Current residue regulations and tolerances (Table I) are, however, based on such "p.p.m. in whole bulk" data and grain containing high residues may

Table I. *Legal and recommended maximum tolerance levels for insecticide residues in stored grain*[a]

Insecticide	Tolerance (p.p.m.)	
	Raw cereals	Cereal products
Allethrin	2	—
bromophos	10	—
DDVP	2	0.3
diazinon	0–0.75	—
Gardona	0.5	—
iodophos	10	—
lindane	0.5	—
malathion	8	0.4–2.0
methoxychlor	0–2	—
piperonyl butoxide	20	20
pyrethrins (natural)	3	1.0
Sevin	2.5	—

[a] See also ROWLANDS (1967), LINDGREN *et al.* (1968), and UN/FAO/WHO (1969).

be "diluted" with untreated or low-residue grain; indeed the "sandwich" technique of treatment was devised with this in mind. It is hoped that some of the results presented in this section, notably those on pesticide dispersion within a given sample, will show that much of the data presented to date is of limited value without reliable assessment of the distribution of pesticide within the grains of such a sample and within the bulk itself.

a) Assessment of residues in practice

With insecticidal treatment of growing crops the residue at harvest is not wanted and is ideally nil, whereas with protective treatment of harvested grain there is deliberate contamination of a cereal food with a pesticidal residue. The weathering and soil leaching effects which decrease residues in field crops do not, of course, apply with postharvest application and the relationship between dose applied and residue obtained should, therefore, be more direct. Thus the following stages can be observed: (1) dose applied, (2) immediate residue recovered = actual dosage, and (3) residue persisting with storage time afterwards. Stage (1) is measured by the operator diluting the concentrate while (2) and (3) are measured (physico-) chemically and/or biologically and are a check on the success of the treatment.

Every chemist and entomologist jointly engaged in considering insecticide residue in grain and attempting to correlate chemical and bioassay figures will be only too familiar with the following anomalies: (a) where good insect control has been achieved in practice but chemical assay of samples withdrawn from the bulk indicate virtually no insecticide present and (b) where grain treated under laboratory conditions to obtain even distribution is toxic for less time than that treated under field conditions. TYLER and ROWLANDS (cited by GREEN 1969) sought an explanation of these anomalies. They found that, in a wheat bulk treated with insecticide that gave good control of both saw-toothed grain beetles and grain weevils, bioassay of 1,000 individual grains showed only five percent of them to be toxic. Similarly, bioassay of 1,000 grains (50 g.) taken from a bulk treated with bromophos showed that 43 grains (two g.) were highly toxic. These 43 grains contained a total of 23 μg. (11 p.p.m.) whereas the rest of the sample contained only eight μg. (0.11 p.p.m.). The overall residue in the total 50-g. sample was 0.6 p.p.m. Under laboratory conditions, complete control of grain beetles in wheat was obtained where one grain in 20 was treated with 40 p.p.m. of malathion, the overall dose being two p.p.m. Clearly, if the presence of a few insecticidal grains in a bulk could exert control, then, depending on the distribution of these grains, chemical and biological assay were likely to disagree at times. Individual wheat grains, taken from a bulk of overall moisture content 14.5 percent, were then treated topically with malathion at rates of ten and 200 p.p.m. At intervals, 50 grains at each rate were bioassayed with grain beetles and chemically assayed. Those dosed at ten p.p.m. were ineffective within four weeks, when only 0.5 p.p.m. of intact malathion remained, whereas those treated at 200 p.p.m. were highly effective over 20 weeks, when 40 p.p.m. remained. If, as found earlier, one grain in 20 containing 40 p.p.m. gives good control it follows that the usually recommended dose of ten p.p.m. would give a longer period of protection if distributed as 200 p.p.m. on five percent of the grains. In practice wheat sprayed (during conveyance at five tons/hour) with emulsion at 1.5 pints/ton to give ten p.p.m. of malathion, contained nine p.p.m. of malathion by chemical assay on 25-g. samples withdrawn from the flow of grain. Actually the range between 548 individual grains assayed from such a 25-g. sample was <0.1 to 424 p.p.m., all grains examined having received insecticide. After 14 weeks storage, the overall residue (25-g. samples) was only 0.5 p.p.m. Had this been evenly distributed it would have been ineffective, but good control of grain beetles continued because the range of insecticide content (based on individual grain assays) was zero to 252 p.p.m.

To express the acceptable contamination level (p.p.m.) in relation to a bulk of whole grains is satisfactory so long as the insecticide is evenly distributed through the bulk and within each grain. However, as GREEN (1969) has pointed out, it has much less relevance if the insecticide is concentrated in part of the seed destined for sale as a separate fraction. If all the insecticide is found in the seed coat, then special attention to residue levels in bran concentrates used in animal feeds is necessary. Certainly insecticides like DDT that accumulate in the germ could result in abnormally high residues in germ extracts for medicinal and other purposes. ROWLANDS (unpublished data) treated wheat grains of 12 percent moisture content with aqueous emulsions of pyrethrins/piperonyl butoxide, DDT, lindane, DDVP, malathion, sumithion, and bromophos. From one to 14 days after treatment, 60 percent of the organophosphorus compounds could be found intact within the endosperm and the other compounds were correspondingly concentrated in the germ. Subsequently the DDVP tended to be translocated to the germ and the pyrethrins were degraded.

b) Routes of penetration into and distribution within a grain[3]

ROWLANDS and HORLER (1967) concluded from experiments with malathion [see section c)] that, even in grain of moisture content

[3] The topics in sections III (a, c, and d) are discussed taking the wheat grain as typical of stored cereal grains. The properties of cereal grains, and wheat in particular, were detailed fully in the earlier paper (*q.v.*). General "structural" references: LINKO (1960), HLYNKA (1964), MACMASTERS (1964), KENT (1966), and TANAKA *et al.* (1967).

as low as seven percent, there is an effective transport system moving the insecticide from one tissue to another.

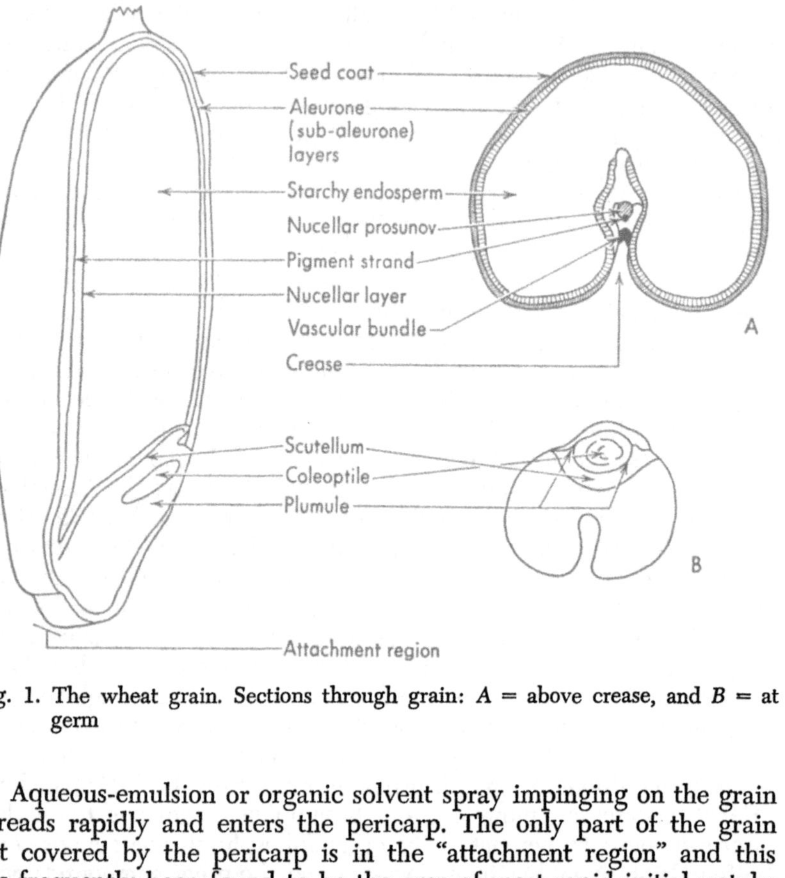

Fig. 1. The wheat grain. Sections through grain: A = above crease, and B = at germ

Aqueous-emulsion or organic solvent spray impinging on the grain spreads rapidly and enters the pericarp. The only part of the grain not covered by the pericarp is in the "attachment region" and this has frequently been found to be the area of most rapid initial uptake (ROWLANDS 1967). In undamaged grains, it is the only route permitting instantaneous entrance and, in addition, the parenchyma tissue of the area contains many intercellular spaces. There is also a capillary system of xylem elements and intercellular spaces leading from the attachment region into the pericarp. The vascular bundle runs through the pericarp tissues from the top of the pigment strand to the base of the kernel and is particularly accessible to spray droplets in the attachment region. The pigment strand *per se* strongly resists penetration by polar solutions as the cutinized walls of the cells are surrounded by lignified layers; in the immature stages, however, the pigment tissues transport water, enzymes, and solutions from the vascular bundle of the pericarp. Rapid uptake may also occur through the myriad intercellular spaces of the intermediate, cross, and tube cells over the lower embryo and along the dorsal surface. In the notch of the crease there are relatively large air spaces where the parenchyma connects with the seedcoat. The partially dry or collapsed cells of the pericarp absorb solutions readily and the cell remnants between the outer and inner tissues also facilitate transport. However, there is, beneath the seed coat and on the walls of the epidermal cells, a thin, relatively water-impermeable cuticle that slows uptake of aqueous media (the nucellar epidermis acts similarly).

In all probability, passage through the pericarp is sufficient to break most emulsion droplets; if not, they are likely to be separated at the water-impermeable cuticular layer which offers less resistance to nonpolar media.

Once through the pericarp the aleurone layer is encountered. This spans the entire length of the starchy endosperm (except where the latter is in contact with the germ) and is closely joined to the nucellar epidermis. The cell walls are striated with small canals between adjacent aleurone and starchy endosperm cells. Each cell has a nucleus surrounded by oil-containing cytoplasm in which many small aleurone grains (proteins) are embedded. These cells contain no gluten and (usually) no starch. The sub-aleurone endosperm contains a single layer of cells and lies alongside the starchy endosperm. Many of the cells contain starch granules and their enzyme and protein content varies according to the species. The walls of these cells are frequently pitted. Aqueous and ionic substances pass easily through the aleurone walls into the cells and are thence transported to and from the starchy endosperm. There is plenty of evidence for the "shunting" of compounds from the starchy embryo to the aleurone layers and to the scutellum.

In addition to the routes already mentioned, when entering through the seedcoat over the embryo, aqueous solutions and polar metabolites solubilised by the cells migrate rapidly into the nucellar layer and into sub-aleurone and aleurone layers before reaching the starchy endosperm. Organic solutions and less polar (fat soluble) compounds (e.g., DDT) are rapidly absorbed into the embryo itself. The micropylar region of the seed coat offers little or no resistance to polar or nonpolar solutions wherefore polar media pass into the embryo via the coleorhiza and root. In addition, the thin-walled cells offer easy passage round the scutellum and thence directly into the endosperm or aleurone layer. In drier grains, nonpolar substances pass directly into the embryonic cavity and thence to the endosperm, whereas the base of the pigment strand, also encountered in this region, is coated with parenchyma tissue and contacts sub-aleurone and scutellum offering easier passageway for aqueous media. Pesticides absorbed into, and degraded and solubilized by, the embryonic axis (e.g., fat-solubles) gain access to the endosperm via the scutellum which lies adjacent. The provascular bundle of the scutellum extends into the upper regions and divides into small branches. It is surrounded by parenchyma cells that comprise the "body" of the scutellum. Very few intercellular spaces are present and the cells each contain nucleus, oil droplets, and aleurone grains.

To and from the aleurone layers, pigment strand, and scutellum there appears to be ready transport of solubilised pesticides and (chiefly) their degradation products via the starchy endosperm. This internal transport may cease under certain physiological or metabolic conditions. For example, direct enzyme inhibition by strong phosphorylators [DDVP, dibrom, phosphine (ROWLANDS 1970 a)], methylating agents such as methyl halides (WINTERINGHAM and HARRISON 1946, ROWLANDS 1969) (see also section V); or by metabolic inhibitors produced by the grains themselves, either from insecticides (ROWLANDS 1966 a) or from natural products (e.g., tocopherols, purothionin, etc.).

The endosperm is a complex lattice of starch-containing cells in a proteinaceous matrix, with a cavity near the nucellar projection; synthetase and phosphatase enzymes are firmly bound to the starch granules and are not readily liberated.

c) *Experimental observations on uptake and the factors involved*

A certain amount of work has been done on the uptake of insecticides by stored grains. In particular attempts have been made to measure the distribution of insecticide within grain by assay of milled or dissected fractions (ROWLANDS 1967). Dry grains can be fairly easily dissected into pericarp(s), endosperm, and germ, roughly corresponding to the bran, flour, and germ of milling practice. This forms a ready, if tedious, method of determining "real" penetration effects at various intervals after treatment. The author does not believe that washing the exterior of the grain is a reliable method of determining the extent of absorption (penetration, uptake). All solvents penetrate grain tissue to some extent and therefore interfere with the assessment of internal and external distribution.

The previous review considered various factors that might affect uptake and, from theoretical approaches, discounted only the age and variety of the grain. For "practical" purposes many more can virtually be discounted. The most influential factors are likely to be the nature of the insecticide and formulation applied, the site of initial application (see above), deposition, or impingement, the environmental temperature, and the moisture content and viability of the grains themselves.

Table II presents data for one "average" grain (on the basis of 100-grain samples) used in the experimental work detailed in this section.

Table II. *Composition data on one "average" wheat grain based on batches of 100 grains each*

	Weight (mg.)			
	Pericarp[a]	Germ	Endosperm	Total (per g.)
Overall weight	0.6	0.12	3.28	4.0
Total protein	0.09	0.03	0.49	0.61
Soluble proteins (including enzymes)				
Albumins	0.04	0.01	0.02	0.07 } 0.15
Globulins	0.04	0.01	0.03	0.08 }
Insoluble protein: (gluten)				
Gliadins	0.01	0.01	0.24	0.26 } 0.47
Glutenins	?	0.01	0.20	0.21 }
Total phosphate content	0.0012[b]	0.008	0.013	0.016[e] (0.4%)
(*ca.* 75% as phytic acid)	0.0006[c]	—	—	—
	0.0004[d]	—	—	—

[a] Includes the aleurone layer.
[b] Aleurone layer (aleurone grains: globoids and aleurins).
[c] Epidermis.
[d] Seed coat (testa).
[e] Total phosphate of protein fraction: gluten 1.5 percent (gliadin 0.7 percent, glutenin 0.7 percent).

Rowlands and Horler (1967) examined in some detail the penetration of malathion-^{14}C, applied topically in 0.1 μl. (640 c.p.m.) of hexane to the back and attachment regions of individual wheat grains. Previous work had indicated that these were the areas of least and most rapid uptake, respectively.

Table III suggests two main effects. Firstly, moisture content has very little effect on the rate of penetration from the back region. Secondly, when applied to the attachment region, there is a marked difference in the amount of malathion reaching the germ, depending on the overall moisture content of the grain. Above 14 percent, very little insecticide enters the germ but below this moisture level approximately 30 percent is found to have reached this tissue. Table IV represents the same determinations on grain of seven percent and 22 percent moisture levels, respectively, but the ^{14}C-contents were separated into hexane-soluble (intact malathion + any malaoxon) and water-soluble (breakdown products) fractions.

There is no difference (Table V) in ^{14}C-distribution, 24 hours after adding the malathion in different solvents. The more volatile solvents evaporated very rapidly after application and are unlikely

Table III. *^{14}C-Content of dissected grain fractions at different times after topical application of ^{14}C-malathion to wheat*

Moisture (%)	^{14}C recovered (%)											
	Pericarp				Germ				Endosperm			
	1 hr.	4 hr.	24 hr.	48 hr.	1 hr.	4 hr.	24 hr.	48 hr.	1 hr.	4 hr.	24 hr.	48 hr.
Back region												
22	100	93	73	45	0	0	2	4	0	7	25	50
17	100	97	76	—	0	0	1	—	0	6	23	—
14	100	97	88	—	0	0	0	—	0	3	12	—
10	100	96	78	—	0	0	1	—	0	4	20	—
7	100	94	77	50	0	0	3	2	0	6	20	47
Attachment region												
22	20	23	23	25	4	4	4	3	76	72	72	71
17	19	26	25	—	4	4	4	—	76	69	70	—
14	7	12	25	—	31	30	28	—	62	57	61	—
10	6	11	22	—	31	29	24	—	64	60	54	—
7	5	10	21	25	22	22	19	12	72	68	60	63

to have any major effect. Dioxane, which is least volatile, did not affect the distribution after 24 hours. Since the volumes used (0.1 μl.) were far in excess of any that would be encountered during "even" spraying in practice, it is unlikely that any organic solvent would affect the later (24 hours plus) distribution of pesticide.

As anticipated, the rate of uptake from the back region was slower than from the attachment region but, despite this difference, the overall picture of distribution is similar (if we except the increased amount of insecticide present in the germ at lower moisture contents). Thus Table VI shows the distribution pattern 48 hours after application to the attachment and seven days after application to the back region. Since, after a few days, the insecticide distributes itself similarly from both positions of application it would seem that, after a week of storage, the internal distribution of malathion is independent of the original site of application. This suggests [cf. section b)] that even in grain of very low moisture content there is an active and effective transport system moving the insecticide and metabolites from one tissue to another.

Table IV. *Separation into intact and breakdown products of ¹⁴C-contents of dissected grain fractions after topical application of ¹⁴C-malathion to wheat*

Fraction	Tissue	Back region: C¹⁴ recovered (%)				Attachment region: C¹⁴ recovered (%)			
		1 hr.	24 hr.	48 hr.	7 days	1 hr.	4 hr.	24 hr.	48 hr.
22% moisture									
Hexane	Pericarp	100	64	42	11	17	22	11	5
	Germ	0	1	2	0	4	3	5	3
	Endosperm	0	26	35	40	78	63	53	48
Aqueous	Pericarp	0	0	3	2	0	1	13	20
	Germ	0	1	2	2	0	1	1	0
	Endosperm	0	8	15	45	0	10	19	23
7% moisture									
Hexane	Pericarp	100	80	49	22	3	9	19	23
	Germ	0	3	2	0	22	20	19	12
	Endosperm	0	17	33	44	75	70	56	59
Aqueous	Pericarp	0	0	1	0	0	0	2	2
	Germ	0	0	1	1	0	0	0	0
	Endosperm	0	0	14	32	0	0	4	4

Table V. *Distribution of ¹⁴C after application of ¹⁴C-malathion in different solvents to the attachment region of 14 percent moisture content wheat*

Tissue	Total ¹⁴C recovered 24 hours after application in different solvents (%)		
	n-Hexane	Carbon tetrachloride	Dioxane
Pericarp	23	26	22
Germ	25	31	27
Endosperm	51	43	51

Table VI. *Distribution of ^{14}C activity two days after application to the attachment region and seven days after application to the back region*

Moisture content (%)	Tissue	Malathion content (%) Attachment (2 days from appl.)	Malathion content (%) Back (7 days from appl.)
22	Pericarp	5	11
	Germ	3	0
	Endosperm	48	40
7	Pericarp	23	22
	Germ	12	0
	Endosperm	59	44

Table VII. *Percent recovery of ^{14}C-dichlorvos*[a] *applied topically to grain*

Tissue	Recovery (%) Back region 1 hr.	24 hr.	48 hr.	7 days	Attachment region 1 hr.	24 hr.	48 hr.	7 days
18% moisture								
Pericarp	100	55	31	3	25	24	25	23
Germ	0	5	7	15	10	10	9	2
Endosperm	0	38	58	74	63	64	66	73
10.6% moisture								
Pericarp	100	68	45	10	5	17	25	28
Germ	0	4	6	21	22	10	7	2
Endosperm	0	23	41	62	72	73	68	68

[a] DDVP + breakdown products (bound and free); any loss of activity due to method or volatization loss.

Table VIII. *Distribution of ^{14}C-activity in whole grains[a] treated topically in the attachment region with labelled dichlorvos*

Extract	Distribution (%)							
	2 days		7 days		14 days		21 days	
	Dichlorvos	Other	Dichlorvos	Other	Dichlorvos	Other	Dichlorvos	Other
Chloroform	24	30	8	45	3	60	NIL	74
Acetone	1	5	2	15	1	15	NIL	7
Aq. saline[b]	NIL	30	NIL	25	NIL	15	NIL	10
Protease digest[c]	NIL	6	NIL	8	NIL	2	NIL	3
Residue[c]	NIL	4	NIL	2	NIL	1	NIL	2

[a] 18 percent moisture content.
[b] As phosphorylated protein.
[c] As phosphorylated protein, amino acids, etc.

Rowlands (1970 a) similarly studied the uptake of DDVP-^{14}C in benzene by the attachment and back regions. He found (Tables VII and VIII) that uptake was very rapid from the attachment region at high moisture level (18 percent). The position found one hour after treatment was practically identical to that found after seven days storage, except for the migration of metabolites from germ to endosperm. Penetration and internal distribution were slower from the back region at either moisture level (10.6 and 18 percent). DDVP applied to the attachment region took longer to penetrate grains of 10.6 percent moisture than those at 18 percent, although at both levels equilibrium was reached after 24 hours. The results agreed with those obtained with malathion (above) in that the internal distribution after two to seven days seemed independent of site of application,

Table IX. *Uptake of ^{14}C-dichlorvos applied topically to the attachment region of wheat grains[a] previously saturated with unlabelled dichlorvos*

Tissue	Stored in sealed vessel (% ^{14}C recovered)				Stored in open vessel (% ^{14}C recovered)			
	1 hr.	24 hr.	48 hr.	7 days	1 hr.	24 hr.	48 hr.	7 days
Pericarp	100	85	85	83	91	43	18	2
Germ	0	2	1	2	0	2	2	2
Endosperm	0	9	9	8	0	8	10	9

[a] 18 percent moisture content.

the main activity being found in the endosperm. Table VIII shows how the overall ^{14}C activity was related to breakdown, with phosphorylation of protein (see later) in each grain during the storage period. During one to two days after treatment (Table VIII, also Fig. 5A) the aqueous-extractable "bound" radioactivity reached a maximum and maintained this level during a seven-day period. Eventually the phosphorylated protein was converted to a more stable, demethylated form and it seems probable that internal distribution of DDVP and its degradation products ceased during the two-to-seven day period when the proteinaceous material was phosphorylated. Results suggesting a similar phenomenon were obtained in earlier work with bromophos (ROWLANDS 1966 a). Table IX shows clearly that if the grain is "saturated" with DDVP and the protein is all phosphorylated, as may pertain in practice when DDVP in solution is atomised and blown through a bulk of grain with an air stream (GREEN and WILKIN 1969), then uptake and internal distribution of DDVP ceases. Under these conditions DDVP is lost only by volatilisation from the grain surface where the surplus remains, only partly absorbed. It is possible that the success of GREEN and WILKIN (1969) at blowing atomised DDVP through the intergranular spaces of a bulk results from the DDVP saturating those grains nearest the point of application and phosphorylating the proteinaceous material therein. Once phosphorylation is complete, the grain ceases to transport the DDVP internally and releases it again into the airflow whence it passes further into the bulk.

Table X. *Uptake of ^{14}C-DDT by wheat grains*

Site of application	Moisture content (%)	Fraction	^{14}C-DDT recovered (%)			
			1 hr.	24 hr.	48 hr.	7 days
Attachment region	12	Pericarp	4	3	3	2
		Germ	48	73	88	91
		Endosperm	45	18	4	2
	17	Pericarp	8	10	6	2
		Germ	41	45	68	88
		Endosperm	48	37	18	3
Back region	12	Pericarp	77	51	30	5
		Germ	2	12	35	78
		Endosperm	21	31	28	10
	17	Pericarp	70	38	12	2
		Germ	4	13	35	86
		Endosperm	20	42	45	3

By contrast, ROWLANDS (1968) studied the uptake of DDT-^{14}C solution topically applied to the back and attachment regions of grains. It was assumed that all the radioactivity recovered was due to DDT since no metabolic degradation of DDT was found during the seven-day observation period. Table X shows that, as with the organophosphorus compounds above, the distribution at seven days is similar at both low (12 percent) and high (17 percent) moisture levels, irrespectively of the initial site of application. The DDT results differed in that the greatest portion of the pesticide was found in the germ tissue. It is quite clear that any DDT metabolism is likely to occur in the germ area.

Interesting though data on the rate of uptake of pesticides may be, it is only the longer-term effects that are of concern to the farmer, vendor, buyer, legislator, and consumer. Undoubtedly speed of uptake in the first 24 hours after treatment has a decided effect upon the metabolism of the compound and its initial rate of degradation but it is the location of residues during later periods of storage that have "practical" relevance. In this connexion, subsequent internal distribution and sites of metabolism within a grain are not influenced by the site of application (impingement of spray, dust particles, etc.),

Table XI. *Summary of results[a] of measuring the rate of uptake of insecticides by wheat grains at 20°C.*

Insecticide applied	Wheat grains (mc[b] < 14.5 %)					Wheat grains (mc[b] > 14.5 %)				
	Organic solvent	E.C.[c]	W.D.S.[d]	Dust	E.C. + S-CMC[e]	Organic solvent	E.C.[c]	W.D.S.[d]	Dust	E.C. + S-CMC[e]
Malathion (deodorised)	M	R	M	M	S–N	R	M	M	S	S
Iodophos	R	M	—	M+	N	R	M	—	M+	N
DDVP	R	R	—	R_v	S	R	M	—	S_v	S
Sumithion	R	M	M	M	N	M	M	S	S	N
Bromophos	R	M	M	M+	N	M	M	S	S+	N

[a] The E.C. and W.D.S. formulations were all commercial products; dusts were commercial except those marked + which were formulated in the laboratory on sterilized (dead) flour; "v" = sorption of DDVP vapour-dust formulation not used. Dose to give ten p.p.m. applied in hexane to jar then grain sample placed one inch above; jar sealed.

R = rapid uptake; 50 percent of applied dose detected intact in endosperm after one to five hours; position largely unchanged thereafter.

M = moderate uptake; 50 percent of applied dose detected intact in endosperm after one to seven days.

S = slow uptake; 50 percent of applied dose detected intact in endosperm after seven to 14 days.

N = slow—no uptake; 50 percent of applied dose still outside after two to three months.

Amounts of water present in aqueous formulations scaled down from 1.5 pints/ton of grain.

[b] mc = moisture content; 14.5 percent is critical.

[c] E.C. = emulsion concentrate (aqueous).

[d] W.D.S. = water-dispersible suspension (aqueous).

[e] E.C. + S-CMC = aqueous emulsion concentrate containing 0.5 percent sodium carboxymethyl cellulose by weight.

the solvent used (if a spray), and the moisture content of the grain, although the latter factor profoundly affects metabolism (see ROWLANDS 1967). Table XI shows the effects that the different formulations used have on the rate of uptake and subsequent location of some common pesticides (cf. data presented by ROWLANDS 1967). It is further evident that the formulation can probably be ignored, except in the case of dusts, which are no longer much used anyway. The nature of the formulation clearly has little effect on the subsequent translocation of pesticide on the basis of longer term results unless an "anti-penetrant", e.g., S-CMC (Table XI, footnote *e*), is added.

d) Locations of enzymes concerned in metabolism

The mechanisms and sites discussed in this section are pertinent only during normal (dry) storage of dormant grain. Once germination is initiated, as in malting or in sprouting of damp grain during storage, completely different circumstances arise and the enzymes are delocalized and/or degraded (DAUSSANT and ABBOTT 1969). Oxidative and hydrolytic enzymes associated with the mitochondria will migrate from the epithelical and parenchyma cells of the scutellum and embryo, from the endosperm, and from the aleurone layer to divers sites of action (cell walls, etc.). During storage of very damp wheat, many enzymes themselves deteriorate rapidly due to an early increase in protease activity (LINKO 1960). The significance of grain moisture content, temperature, etc. on enzymic activity has been discussed in the earlier review.

The main enzymic activity in protein extracts of wheat grains has been found in the globulin fraction (MEREDITH and WREN 1966, ROWLANDS 1970 a) and, at tissue and cellular levels, in the aleurone grains and cytoplasm of the aleurone and germ parenchyma cells (ROWLANDS 1966 b). An insecticide penetrating the newly-harvested wheat grain will have some slight opportunity of being attacked (oxidation, hydrolysis, decarboxylation) by the field-fungal hyphae invariably present between the outer and inner pericarps or by the phenoloxidase-type systems, concerned in the last stages of ripening, in the vascular bundle of the pericarp and the pigment strand. As a result of rapid penetration to the aleurone layer, endosperm, and germ, these processes are of little importance: in any case they are active for only two to three weeks after harvest. The aleurone layer and germ are the major areas of insecticide metabolism within the grain. Enzymes effecting hydrolysis and, to a lesser degree, decarboxylation and oxidation of insecticides, have been identified in the aleurone layer (ROWLANDS 1967). There is no evidence of anaerobic metabolic activity in the aleurone cells and it is presumed that little would occur in this region during prolonged storage should anaerobic respiration of the grain supervene within the bulk. The parenchyma cells of the scutellum and embryonic axis of the germ are actively concerned in the metabolism of insecticide and more so, in the case of oxidation and decarboxylation, than the aleurone layer. In addition, these germ tissues seem to be responsible for the switch to anaerobic respiration of the grain during long-term storage (years), particularly the coleoptile region of the embryonic axis. Since there is facile transport of solubilized pesticide and metabolites to and from the "metabolic" regions by the endosperm, some of the less labile metabolites (e.g., the substituted phenols) may accumulate there (ROWLANDS 1966 a), while others, e.g., phosphoric acids, aliphatic acids, alcohols, etc., may be incorporated into the normal metabolism of the grain. SOBOLEV and RODIONOVA (1966) found rapid incorporation of "foreign" ^{32}P into ATP and phytin by aleurone grains; NOVOTEL'NOV *et al.* (1966) studied incorporation of organic acids into pericarp tissues and MITSUDA *et al.* (1968) studied the volatile alcoholic and carbonyl compounds.

1. Oxidation. The phenoloxidase-type enzymes of freshly harvested wheat have been shown to oxidise phosphorothionates to phosphorothiolates and phosphates (P=S→P=O), and substituted (chlorinated) cresols to di- and trihydroxy compounds *in vivo* (ROWLANDS 1966 b, ROWLANDS *et al.* 1968, ROWLANDS 1970 b). Whereas crude *in vitro* preparations were successful in achieving this, a highly purified tyrosinase isolated from freshly-harvested or ripening wheat seed coats did not do so (ROWLANDS *et al.* 1968). Similar failure was observed with a commercial (mushroom) tyrosinase although both laboratory and commercial preparations were active towards *p*-cresol.

ROWLANDS (1966 b) observed slight phosphorothionate oxidation in the germ tissues of freshly-harvested wheat grains but did not investigate the mechanisms involved. However, during a study of the role of lipoxidases (peroxidation of linoleic acid) *in vitro* under aerobic and anaerobic conditions (partly described by ROWLANDS (1968), ROWLANDS (1970 b) found that the fat-peroxide systems, using either a wheat germ lipoxidase or commercial catalase or peroxidase, would concurrently oxidise P=S to P=O and that preparations from older grains (> eight months) were consistently more effective at this linked thiono-thiolo conversion than were those from freshly-harvested grain. This finding was also borne out *in vivo*. Moreover, LITTLE and O'BRIEN (1968) reported oxidation of thiol groups by lipid peroxides and found that these reactions were catalysed by haemin compounds which are generally believed to decompose lipid peroxides into free radicals. The small amount of phosphorothionate activation so far observed during storage of treated grain (always < ten percent of applied dose (ROWLANDS 1967) is probably because most of this work has concerned fresh grain. It now seems that older grain may achieve more P=S to P=O conversion. The same peroxidizing systems converted pyrethrin I and allethrin into some four or five products *in vitro* but were not active *in vivo*, considerable hydrolytic degradation preferentially occurring. The aerobic systems were not active against DDT (in terms of "oxidation products", e.g., Kelthane, DDA, or ring-hydroxylation) either *in vitro* or *in vivo* but did produce trace amounts of DDE. Anaerobically they dechlorinated DDT to DDD (see below). Under anaerobic conditions the systems were ineffective at oxidizing phosphorothionates and pyrethrins but some hydrolysis of both classes of compound occurred.

It seems reasonable to suppose that oxidation of organophosphorus and pyrethroid compounds in the germ area will involve the electron-transport systems generally (possibly the microsomal cytochrome P 450) and be linked, in particular, with the normal areobic peroxidation of unsaturated fats (Gardner and Inglett 1968).

Theoretical mechanisms of aromatic hydroxylations via intermediate epi-peroxides have been discussed by Soloway (1966).

2. Hydrolysis. Hydrolases have been located in the aleurone layer and in the parenchyma region of the scutellum as well as (transiently) in fungal hyphae between the pericarps (Rowlands 1967). These hydrolases comprise the specific and nonspecific lipases, phosphatases, and esterases, many or all of which may be isozymes of the same enzyme. Park and Koshland (1958) have shown that 3-phosphoglyceraldehyde dehydrogenase, in the absence of NAD, showed esterase activity. On blocking the sulphydryl groups with iodoacetic acid, the esterase activity disappeared in favour of phosphatase activity. Many organophosphorothiolates and phosphates (P=O compounds) inhibit these esterases (discussed more fully in section V) and DDVP, in particular, forms phosphorylated protein derivatives that are mainly water-soluble (Rowlands 1970 a).

Allethrin (and pyrethrin I) were degraded to chrysanthemic acids and parent alcohols by crude *in vitro* hydrolase preparations from germ and aleurone layer and also (see above) by certain lipoxidase preparations.

3. Decarboxylation. Most of the activity located in the grain has been associated with the germ or parasitic microflora but some has been observed in the aleurone layer.

4. Dehalogenation. Rowlands (1968) found very slight dehydrochlorination to DDE of DDT in stored wheat grains under aerobic conditions but once the intergranular air was consumed (or, rather, when these conditions were simulated) and anaerobic conditions prevailed then reductive dechlorination to DDD took place. These reactions were localised in the parenchyma cells of the scutellum and embryo, seemed to be linked with the anaerobic peroxidation of unsaturated fats by iron-porphyrin compounds and were enhanced by the presence of carotenoid compounds. Natural inhibitors (e.g., purothionins) present in the germ may regulate the mechanisms. The true lipoxidase system (which was more effective at peroxidizing unsaturated fats under aerobic conditions and also oxidised phosphorothionates and pyrethrins) alone, or in the presence of carotene, had no effect on DDT anaerobically or aerobically. It would work only in the presence of such commercial iron-prophyrins as catalase, peroxidase and cytochrome C or an extract of prophyrins from wheat. CASTRO (1964) and MISKUS *et al.* (1965) found that dilute solutions of reduced iron-prophyrin (Fe^{2+}) complexes were oxidized by DDT *in vitro* to corresponding haemins (Fe^{3+}-halide) while DDT was reduced to DDD. TAPPEL (1953) reported catalytic anaerobic oxidation of unsaturated fats by traces of iron-prophyrin oxidases, similar to the known aerobic role of lipoxidase (GARDNER and INGLETT 1968). HAWTHORN and TODD (1955) and GARDNER and INGLETT (1968) showed that, in the wheat germ, compounds such as catalase could play an important role in the oxidation of fats, the germ being richer in catalase and peroxidase than in true lipoxidase. The oxidation of carotene(s), xanthophylls, and other such compounds is due to another coupled reaction with the peroxidized linoleic acid. Since there is little evidence for a direct link between lipid peroxidation and DDT formation, unsaturated fatty acids may facilitate the reaction by solubilizing DDT. If this is so, carotene might be protecting the iron-porphyrin compounds from the action of lipid peroxides.

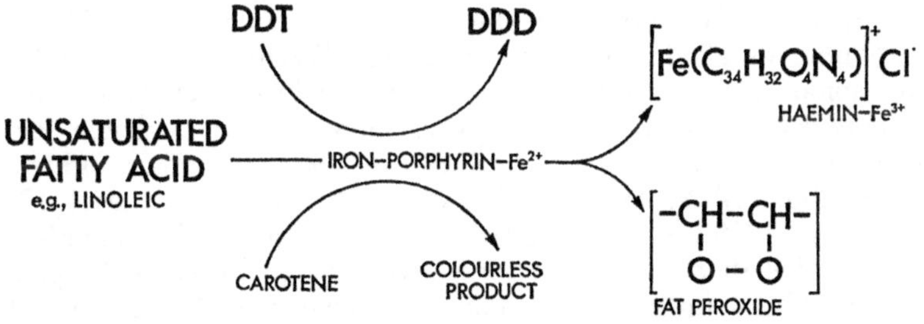

Fig. 2. Suggested complex of reactions concerned in anaerobic DDT metabolism by wheat germs

Figure 2 shows a suggested complex for the reactions concerned in reductive dechlorination of DDT by wheat germ under anaerobic conditions. It is not yet known to what extent these systems may be involved in other de(hydro)-halogenations, for example in degrading lindane or aromatic halogenated compounds (e.g., "leaving groups" from organophosphates). Tentative results indicating anerobic dechlorination of both lindane and methoxychlor were obtained by ROWLANDS (1970 b) but these compounds were little affected by aerobic systems.

Dehydrochlorination of dipterex to DDVP by insects, plants, and animals has been reported by SUGIYAMA and SHIGEMATSU (1969) and by BULL and RIDGWAY (1969).

SCOTT (1967) discusses the utilization by plant cells of de(hydro)-halogenation products derived from "foreign" materials.

5. Dealkylation (desalkylation). *O*-dealkylation has frequently

$$(CH_3O)_2P(=O){-}OR \rightarrow \begin{matrix} CH_3O \\ \quad\diagdown \\ \quad\quad P(=O){-}OR \\ \quad\diagup \\ HO \end{matrix}$$

been identified as a degradation pathway, usually [but not invariably (AHARONI and O'BRIEN 1968)] detoxication, of organophosphorus (and other) compounds in mammals, insects, plants, and microorganisms. There is sometimes clear specificity of attack limited to methyl, rather than ethyl, phosphate esters. Properties of the dealkylation systems of mammals and insects have been examined *in vitro* by FUKAMI and SHISHIDO (1966) and by HOLLINGWORTH (1969). The mechanisms have often proved to be "hydrolytic" though *in vitro* co-factor requirements suggest that in many cases microsomal "oxidases" may be implicated: the predominant mammalian mechanism *in vitro* seems to be transfer of an alkyl group to glutathione (HOLLINGWORTH 1969); the functioning of alkyl phosphate esters as alkylating agents could explain some of the side-effects, and synergistic effects, observable in organophosphate toxicosis. However, there is also evidence for involvement of phosphatase (hydrolytic) attack in mammals and insects (MAIN 1960, MATSUMURA and SAKAI 1968, HOLLINGWORTH 1969); ROWLANDS (1966 a) concluded that demethylation of bromophos by wheat grains *in vitro* and *in vivo* was effected by acid-phosphatase type hydrolyses rather than oxidative mechanisms. The crude nature of his "*in vitro*" preparations, however, did not preclude involvement of the oxidative system. No further work has been carried out to elucidate this mechanism directly but ROWLANDS (1970 a) found that a commercial acid-phosphatase preparation (from yeast) demethylated the water-soluble dimethyl phosphoryl-protein complex, formed by DDVP with wheat protein, to a more stable monomethyl phosphoryl derivative. ROWLANDS (1969) found that methylating wheat grains with a large dose of methyl bromide fumigant interfered with the demethylation of sumithion later applied to the grain during storage.

The relevance of alkylating agent (especially fumigant) residues and their effects on insecticide metabolism within the grain will be considered under "synergism" in section V.

S-Dealkylation (which might be a possible degradation pathway of certain minor insecticide products such as S-alkyl isomers $\genfrac{}{}{0pt}{}{RO}{RS}\!\!>\!P(=O)-OR$ or decarboxylated carboxyesterase products, e.g., $(RO)_2P(=S)-SCH_3$ is much less likely to occur. Mercaptans $(-SH)$ seems to have considerable affinity for alkylation and, as implied above, may well play biochemical (as with glutathione) or chemical [as with dimethyl phosphorodithioate, $(CH_3O)_2P(=S)SH$] roles as receptors for alkyl groups removed from *O,O*-dialkyl (thio) phosphates.

e) Grain enzymes, general literature review

This section merely comprises references to recent papers on grain enzymes in general.

An extremely important paper was Linko's (1960) investigation into the water-content and metabolism of wheat during short storage and germination. It dealt with, and updated, many of the topics covered in the classic review of Kretovitch (1945) and with glutamic acid decarboxylase in particular.

Altschul *et al.* (1966) gave a broad review of grain enzymes while considering cereal proteins.

Enzymes or isozymes studied that are of particular significance in insecticide metabolism include: esterases (Stauffer and Glass 1966, Rothe *et al.* 1967), isozymes of lipoxidase (Guss *et al.* 1967), peroxidase (Gardner and Inglett 1968), and the coupled reaction with carotene (Kies *et al.* 1969) and porphyrins and phenoloxidases (Honold and Stahmann 1968). Thermal complexes formed by starch with organic substances have been reviewed by Kuge and Takeo (1967).

More generally, dormancy in cereals and control and location of enzymic activity during dormancy was the subject of reviews by Varner (1965), Major and Roberts (1968), and Roberts (1969) while Dennis and Coultate (1966) discussed phosphofructokinase as a regulatory mechanism of plant metabolism. Lipids (Pomeranz 1966, Wolfe 1966, Gardner and Inglett 1968, Rooney 1968, Weber 1968), proteins (Altschul *et al.* 1966, Ludwiczak *et al.* 1967, Zubaidov 1968), and lipoprotein complexes (Rohrlich and Niederaver 1967) of cereal grains have been well discussed as have the changes in these factors during long-term storage (Pixton and Hill 1967). Macko *et al.* (1967) have dealt with multiple enzyme systems, Nip and Burns (1968) with pigments, and Thompson (1967) with sulphur metabolism.

Schemes for respiratory electron-transport systems of plants generally have been proposed, and a great deal of evidence reviewed, by LIEBERMAN and BAKER (1965) and GREEN and SILMAN (1967), while of cereal grain oxidases, the dehydrogenases have received particular attention (LINKO 1960, HONOLD *et al.* 1966 and 1967, HONOLD and STAHMANN 1968, ROWSELL and ALI KHAN 1966). Studies on inositol and starch synthetases (KURASAWA *et al.* 1967, MURATA *et al.* 1968), amylases (MEREDITH 1966, KRUGER and TKACHUK 1968), cellulase (KATZ 1967), peptidase (PRENTICE *et al.* 1967), and methionine methyl transferase (KARR *et al.* 1967) have been reported.

The enzymic activity of the parasitic microflora (CLARKE 1968, L'VOVA 1968), enzymes of *Alternaria* spp [a long series of papers, see VERMA (1964) for references] and *Aspergillus* spp. (NEALSON and GARBER 1967) have received attention. AYERST (1969) reported on the moisture and temperature limits of growth for 12 species of *Aspergillus,* three *Penicillium,* and *Stachybotrys atra.* As in the past, mechanisms of aromatic hydroxylation have continued to be investigated (RIBBONS 1965, SOLOWAY 1966, EVANS 1967, FONKEN *et al.* 1967, SLOANE and HEINEMANN 1967). WILSON (1967) has reviewed the toxins produced in foodstuffs by fungi.

The following naturally-occurring enzyme and growth inhibitors and regulators have been studied or reported on (see in greater detail in section V): trypsin inhibitors (LEARMOUTH and WOOD 1960, SHYAMALA 1961), antioxidants in oats [a long series of papers, see DANIELS and MARTIN (1968) for references], dormins from xanthophylls (TAYLOR and SMITH 1967), giberellins (MACLEOD and PALMER 1967), phytic acid (CAWLEY and MITCHELL 1968), purothionins (FISHER *et al.* 1968, REDMAN and FISHER 1969), and tocopherols (DRAPER 1968, HALL and LAIDMAN 1968, SLOVER *et al.* 1968, WITTING 1968, HERTING and DRURY 1969).

IV. Insecticide residues and metabolism in stored grain

The extensive review of LINDGREN *et al.* (1968) included selected data on postharvest pesticide residues and persistence in cereal grains. Since that topical survey, a further mass of results has collected. However, this section will deal mainly with positive and tentative attempts at identification of the metabolic products themselves, unlike that in the previous review which included a good deal of speculation about residue work not specifically undertaken to investigate metabolism.

Attenuation of residues on seeds and foodstuffs has been reviewed by KOHN (1967 and 1970) and STREET (1969).

a) The pyrethroids

ROWLANDS (1968b) made a preliminary study of (+)-dl-*trans*-allethrin degradation when applied at ten p.p.m. to wheat grain of

15 percent moisture content. Assay of residues by gas (GLC) and thin-layer chromatography (TLC), allied with infrared and mass spectroscopy, showed that the half-life at 20°C. was approximately four weeks and suggested hydrolytic scission at the chrysanthemic ester linkage as the main degradative pathway. Joint application of ten or 100 p.p.m. of piperonyl butoxide with the ten p.p.m. of allethrin did not prolong the intact residual life of this insecticide.

More intensive work (ROWLANDS 1970 b) with synthetic d-*trans*-d-pyrethrin I and d-*trans*-d-allethrin resolved the hydrolytic products qualitatively into a common major metabolite, d-*trans*-chrysanthemic acid, two lesser unidentified (acidic) compounds, and the respective keto-alcohols d-pyrethrolone and d-allethrolone. These were the only products obtained on grain *in vivo* (as it were) and the presence of piperonyl butoxide, tropital (piprotal), sesamex, or BHT had no effect on the qualitative "spectra" of metabolites obtained. The respective half-lives established for pyrethrin I and allethrin in grain of 15 percent moisture content at 20°C. by this investigation were three and four weeks.

As mentioned earlier [section III d) 1], additional products were obtained *in vitro,* using fat peroxidation systems (grain lipoxidase and commercial catalase and peroxidase preparations) under aerobic conditions. The grain system was most effective when prepared from older wheat (> eight months). The same hydrolytic products were obtained as *in vivo* but there were also five further (presumably oxidative) metabolites, three of which had the ester linkage intact and production of which was inhibited by piperonyl butoxide, tropital, sesamex, or BHT.

The elegant work of YAMAMOTO and CASIDA (1966) and YAMAMOTO (1969), with a housefly enzyme system *in vitro,* provided an indication as to the nature of some of the additional products. They had shown that ^{14}C-pyrethrin I and ^{14}C-allethrin were oxidized at the 1-methyl group of the isobutenyl moiety of the chrysanthemic acid component, to yield *O*-demethyl pyrethroid II analogues. They found similar conversion of allethrin *in vivo* when injected into houseflies. In addition they established *in vitro* oxidation of analogous groups of dimethrin and phthalthrin, thus suggesting a detoxifying pathway of general importance for all esters of chrysanthemic acid. They thought that absence of the isobutenyl group in other pyrethroid esters (pyrethrin II, cinerin II, etc.) might explain the markedly lower synergism between sesamex and these compounds. CASIDA *et al.* (1966) obtained *in vitro* hydroxylation of ^{14}C-methylenedioxyphenyl synergists by the same enzyme preparations.

Using the grain (and commercial) lipoxidase systems under aerobic conditions the following products were obtained from allethrin *in vitro* (ROWLANDS 1970 b): demethyl allethrin II, an unidentified (neutral) compound with intact chrysanthemic acid linkage, and a

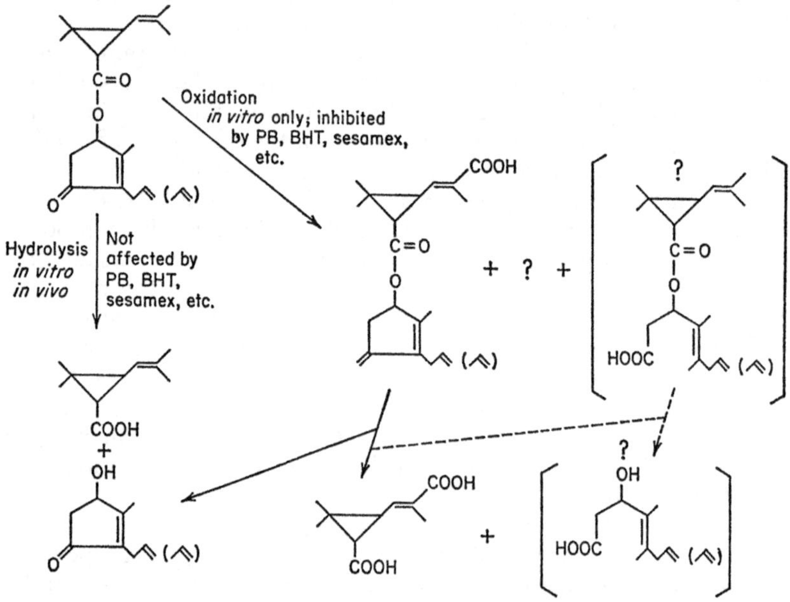

Fig. 3. Pyrethroid metabolism in grain

similar but acidic compound that infrared and mass spectroscopy suggested might have resulted from oxidation of the keto group in the alcoholic moiety. All the "hydrolytic" products previously obtained *in vivo* were detected and also chrysanthemum dicarboxylic acid and a neutral product lacking the chrysanthemic acid moiety. Piperonyl butoxide, tropital, sesamex, and BHT all prevented or decreased production of the oxidative metabolites.

The anaerobic lipoxidase system (as used to dechlorinate DDT) did not produce any of these oxidation products although small amounts of the *in vivo* hydrolytic metabolites were obtained. None of the oxidation products were obtained when allethrin or pyrethrin I were applied to stored whole wheat grains. Figure 3 shows some of the postulated metabolic pathways for pyrethroid I-type insecticides in grain.

In practice, it seems, little or no oxidative metabolism occurs of chrysanthemic esters applied to grain, hydrolytic degradation being the usual pathway. This is not surprising since pyrethroids applied to grain have little opportunity to be attacked in passing through the seedcoat and, once inside, the "opportunity factor" and general lability probably ensure rapid hydrolysis rather than oxidation. In any case oxidation has only been demonstrated *in vitro* and does not appear to be very active in grain less than eight months old.

Normal pyrethrum synergists and antioxidant-stabilisers do not therefore prolong the intact residual life of such compounds.[4] So far no attempts have been made to identify or to separate the metabolites and complex degradation products that result from application to grain of the mixture of natural pyrethrum (the pyrethrins, cinerins, jasmolins, etc.).

b) *Organochlorine compounds*

1. DDT and methoxychlor. ROWLANDS and SHARMA (1968) stored damp and dry wheat grain treated with DDT under air, nitrogen, or carbon dioxide. They found no apparent breakdown of DDT in any sample during one month when most of the applied dose was in the germ. During subsequent storage for three months, ten to 20 percent loss of DDT occurred in both damp and dry samples under N_2 and CO_2 and in damp samples stored under air. Qualitative assay (TLC) of the metabolites produced indicated only DDD (dechlorination) in the grain under N_2 or CO_2; in the damper sample stored under air, DDE (dehydrochlorination) was the main product with trace amounts of DDD present. During the three-month period the damp grains became visually mouldy and after four months the germs of the grain stored under CO_2 blackened and died. ROWLANDS (1968 a) confirmed these findings in that DDT applied to stored wheat grains underwent slight dehydrochlorination to DDE while aerobic conditions existed but subsequently degradated via reductive anaerobic dechlorination to DDD, which would still be insecticidal. Experimental evidence located these reactions in parenchyma cells of the germ and linked the anaerobic mechanism with the peroxidation of unsaturated fat by iron-prophyrin compounds [section III d) 4, Fig. 2]. *In vitro* the conversion of DDT to DDD was enhanced by carotene and was not inhibited by the organophosphates malaoxon and DDVP, nor by the insecticide synergists SKF 525A and sesamex or antioxidants such as BHT, BHA, or hydroquinone. This lack of effect may have been due to the anaerobic conditions since these compounds inhibit aerobic lipoxidase. Of natural (lipoxidase) inhibitors, purothionin(s) and α-tocopherol, only purothionin was effective anaerobically *in vitro* and decreased the decolouration of carotene and degradation of DDT. No evidence was obtained of any "oxidative" metabolism to Kelthane, DDA, or ring-hydroxy products either aerobically or anaerobically. However such products might arise at times in mouldy wheat. ENGST and KUJAWA (1967) found that fungi (*Fusarium* spp.) degraded DDT to DDA and DDEE in addition to

[4] It should be remembered that "bioassay" with insects may give an erroneous picture if residues of synergist potentiate the toxicity of small (trace) residues of the pyrethroids left in the grain. These effects are less (or not) observable with the synthetic pyrethroids, some of which are difficult to synergise.

producing DDD and DDE. ROWLANDS and LLOYD (1969) found that insect frass (microorganisms present in?) from grain weevils breeding in stored wheat treated with DDT produced DDA as well as DDE and DDD.

Physiological effects of DDT (KAGAN *et al.* 1969) or metabolic products thereof, particularly on photophosphorylations in barley plants have been observed (LAWLER and ROGERS 1967 a and b), whereas earlier results on the lack of phytotoxicity to grain embryos have been confirmed (ZAKL *et al.* 1966).

Methoxychlor applied to stored wheat underwent slight dechlorination to methoxy-DDD but seemingly was not attacked by the aerobic system either *in vivo* or *in vitro* (ROWLANDS 1970 b).

2. Lindane. Despite the continued use of lindane for application to stored grain, there seems to have been no further work, other than reviewed earlier, on its metabolic fate therein. However ROWLANDS (1970 b) found that lindane was slightly dechlorinated in whole grains, and by the anaerobic lipoxidase system *in vitro*, to γ-pentachlorocyclohexane (PCCH) and tetrachlorocyclohexane, less-toxic products; no trichlorobenzenes (TCB) were detected. Lindane was apparently little affected by the grain systems under aerobic conditions though traces of TCBs were obtained; it should be stressed that these observations are only tentative. It is not clear to what extent the DDT and lindane de(hydro)chlorination systems in grain may overlap.[5]

The lack of phytotoxicity of lindane applied to stored grain [see ROWLANDS (1967) for references] may be taken as an indication that little or no aerobic degradation to the injurious TCB occurs. As with DDT, more pronounced metabolism (ring fission and ultimately to CO_2) might be effected by the parasitic fungi of grain (MACRAE *et al.* 1967 and 1969, YULE *et al.* 1967). SUGIMOTO (1966) followed the slight translocation of lindane from soil to the maturing grains of rice plants and determined the amounts subsequently concentrated in bran and endosperm.

c) *Organophosphorus compounds*

The metabolic fate of several of the organophosphorus insecticides used, or of potential use (e.g., malathion, sumithion, ronnel, bromophos), on stored grain was discussed fully in the previous review. These will not be detailed further, therefore, unless new or interesting factors have come to light. Figure 4 shows the general pathways suggested for the metabolism of organophosphorus insecticides in stored grain.

---

[5] ISHIDA (1968) found the DDT-dehydrochlorinase and BHC-metabolizing systems in houseflies to be very similar.

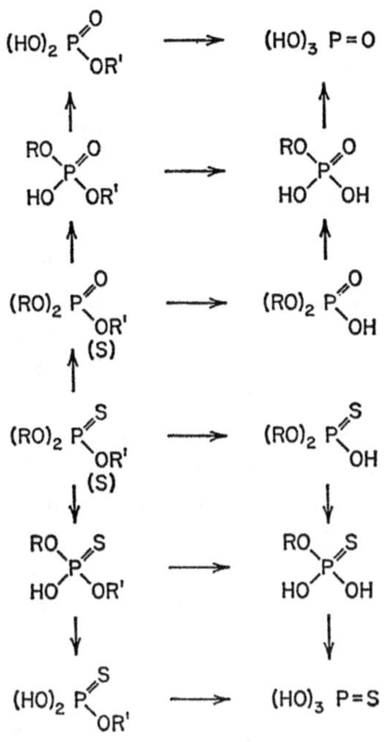

Fig. 4. Metabolism of phosphates

1. Phosphates. Rowlands (1970 a) has shown that ^{14}C-labelled DDVP broke down rapidly on wheat grain to give mainly dimethyl phosphate and phosphorylated protein derivatives. Lesser amounts of desmethyl-DDVP, monomethyl phosphate, and phosphoric acid were found. Moisture content of the grain did not have much effect on the degradative rate, which was rapid even in dry grain, but damper wheat produced more of the water-soluble phosphorylated proteins at the expense of other products. Investigation of these proteins, after Sephadex gel fractionation, suggested that there was little molecular weight difference between the extractives of DDVP-treated and untreated wheats and that DDVP probably dimethyl phosphorylated available hydroxy groups of the protein molecule, e.g., PROTEIN—O—P(=O)$(OMe)_2$. This complex underwent rapid desmethylation during storage to give a more stable form, PROTEIN—O—P(=O)(OMe)(OH). The initial complex possibly consists of dimethyl phosphoryl-derivatives of the constituent amino-acids serine, threonine, and histidine, since these were the only amino-acids to be phosphorylated *in vitro*. Digests of the complexes extracted from treated grain yielded mono- and dimethyl phosphoryl derivatives of amino acids and peptides which were then degraded more positively to give phosphoryl serines. Acid phosphatase slowly attacked these phosphorylated serines *in vitro* to yield mono- and dimethyl phosphates. Serine appeared to be the only amino acid phosphorylated *in vivo;* products of histidine and threonine were not detected in grain digests.

Among other phosphates examined only dibrom was as effective at producing water-soluble phosphoryl derivatives from wheat protein; the fumigant phosphine seems to produce similar effects, though the nature of the complex has not yet been investigated (BERCK 1968); but see ROHRLICH and MEUSER (1969). Malaoxon, sumioxon, triphenyl phosphate, and Gardona were also lesser phosphorylators of wheat protein. This is not surprising in view of their known inhibitory properties against enzymes such as cholinesterase and carboxyesterase. Phosphorothionates and dithioates did not show any ability to phosphorylate protein.

None of the phosphorylated protein fractions from DDVP-treated wheat were toxic to adult *Oryzaephilus surinamensis,* but germination was slightly impaired at the ten p.p.m. level and seriously when saturated with DDVP, as was the enzymic activity generally. At the ten p.p.m. level, enzyme impairment was limited to esterase-types and DDVP might thus prolong the residual lives of some hydrolysable insecticides if present in the grain; above 20 p.p.m., some impairment of phosphorothioate oxidation was observed [see section V b) 4 below]. DDVP appeared to compete with other phosphates and phosphorothioates for degrading enzymes. Figures 5 A and B show the quantitative degradation rates of DDVP in grain and accumulation of metabolic products. ROWLANDS (1970 c) showed that there was more rapid breakdown of DDVP on barley than on wheat when both were stored at 12 percent moisture content and 20°C. for seven days.

DDVP residues might possibly arise from dehydrochlorination of the phosphonate, dipterex (trichlorfon). BULL and RIDGWAY (1969) have shown this pathway to occur in plants. Other such pathways were considered in the earlier review. Dipterex is a powerful alkylating agent [see HOLLINGWORTH (1969) for references].

Gardona (SD8447) metabolism in grains has not been studied in any detail. ROWLANDS (1968 b) did not find it to be very stable on damp wheat, ten p.p.m. applied to grain of 18 percent moisture broke down approximately 20 percent in two to four weeks to give mainly the desemethyl derivative. Quantitative GLC determination was impeded by tetrachloroacetophenone, a contaminant and pyrolysis product of the crystalline insecticide standard. Later work with ^{14}C-Gardona on wheat grain showed additional degradation to TCA,

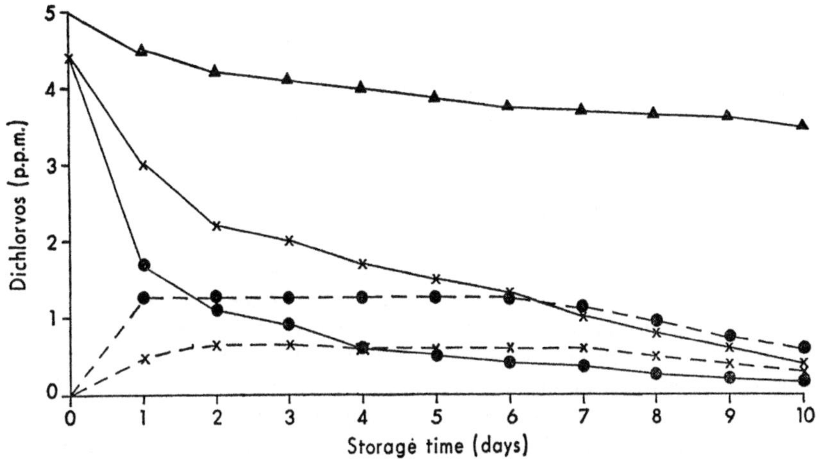

Fig. 5A. Degradation of ^{14}C-DDVP on stored wheat grain at 20°C., no vapour loss: ▲——▲ autoclaved wheat (18 percent moisture), ×——× chloroform solubles (intact DDVP) from ten percent moisture wheat, ×- - -× water-soluble protein complex in ten percent moisture wheat, ●——● chloroform solubles in 18 percent moisture wheat, and ●- - -● water-soluble protein complex in 18 percent moisture wheat

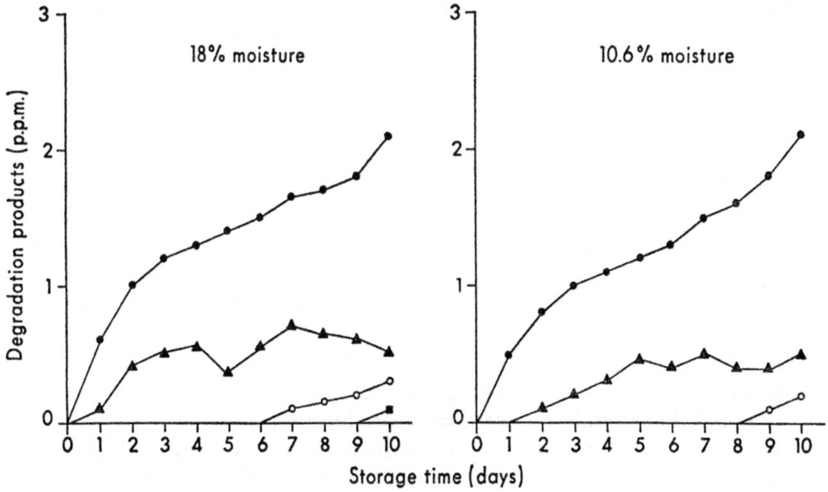

Fig. 5B. Degradation products of DDVP on wheat: ●——● dimethyl phosphate, ▲——▲ desmethyl DDVP, ○——○ monomethyl phosphate, and ■——■ phosphoric acid(s)

TCC, and TCE; however desmethyl Gardona was the main product. As pointed out in section II, several of these products are more toxic to mammals than the parent compound. BEYNON and WRIGHT (1969) studied the breakdown of Gardona in soils by cabbage and rice plants and by apple foliage and fruit. In addition to the products mentioned above they found the bisdesmethyl derivative and conjugates of the alcohols, TCA and TCC.

2. Phosphorothionates and phosphorodithioates. Little further work has been done on the metabolism of compounds with aryloxy- or arylthio-phosphate side chains. No evidence was obtained by ROWLANDS (1970 b) of reduction at the nitro group of sumithion to give amino-sumithion or derivatives (YASUNO *et al.* 1965, MIYAMOTO *et al.* 1966) in stored grains under anaerobic conditions, but there was some suggestion that the endemic fungi, or microorganisms present in insect frass, might affect such conversion under anaerobic conditions, as traces of amino-cresols were found in samples from heavily infested wheat grain that had been treated with sumithion but which subsequently heated and became mouldy. HORLER *et al.* (1969) observed that degradation of sumithion by liquid cultures of the grain fungus *Alternaria tenuis* was slower than that of bromophos or malathion. The fungistat 8-hydroxyquinoline had no effect on the breakdown of sumithion, malathion, or bromophos in grain of 12 percent moisture content (HORLER *et al.* 1968); however, little or no fungal development could be expected at this moisture level.

HORLER *et al.* (1968 and 1969) determined the degradation of bromophos by liquid cultures of *A. tenuis* and observed the half-life to be 24 hours at pH 5. They found dichlorobromophenol to be an important phosphatase hydrolysis degradation product that inhibited further *in vivo* breakdown of bromophos but had no effect on the exogenous metabolism of the fungus. STENERSEN (1969 a) studied ^{32}P-bromophos metabolism by certain microorganisms including the grain fungus *A. tenuis.* The main metabolite detected seemed to be the bisdesmethyl derivative (STENERSEN 1969 a and b) which he considered to derive directly from bromophos and not via the desmethyl compound.

Preliminary (unpublished) investigations on the metabolism of iodophos by stored grain have indicated that pathways similar to those of bromophos and ronnel are involved, namely, desmethylation, lesser hydrolysis to dimethyl phosphorothionate and dichloroiodophenol and low levels of the P=O analogue (iodoxon). Breakdown in wheat at both high and low moisture levels is slower than with bromophos and ronnel, however, possibly due to the stabilizing effect on the molecule of the *p*-ring substituent; iodo- being less electrophilic than bromo- or chloro-. This slower breakdown, which is more pronounced at lower moisture levels, could mean increased "opportunity" for oxidation by the germ lipoxidase complex. WILKIN *et al.* (1970) found that iodophos applied to barley of 14 to 27 percent moisture content that was heating badly in store degraded from five p.p.m. to three p.p.m. within seven days and to 2.4 p.p.m. within 14 days, but these were extreme conditions.

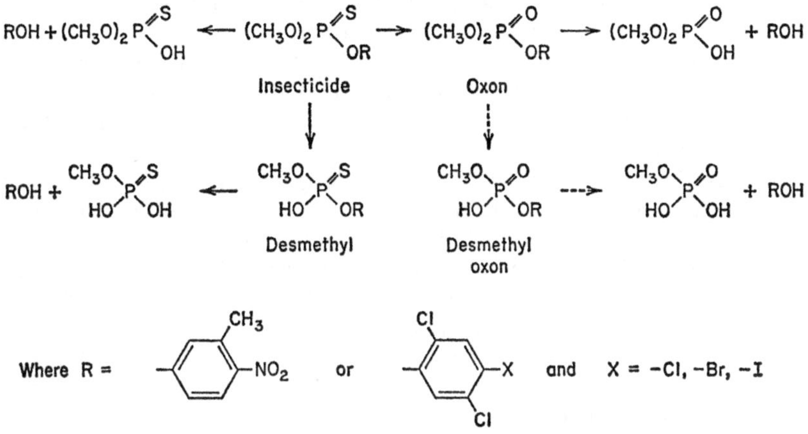

Fig. 6. Metabolism of sumithion, Ronnel, bromophos, or iodophos in grain

An interesting "metabolic" advance in studies with malathion[6] has been the work of DAUTERMAN's group (CHEN *et al.* 1969) who have resolved the structure of the "biologically-produced" malathion monoacid (carboxyesterase product) as the α-isomer, namely

$$(CH_3O)_2P(=S)\underset{\displaystyle |\atop \displaystyle CH_2COOC_2H_5}{SCHCOOH}$$

Although desmethylation of malathion does not seem to occur in grain, whereas other carboxyester phosphorodithioates such as dimethoate and Cidial are so degraded, it is of interest that MATSUMURA and SAKI (1968) separated the malathion desmethylase of the American cockroach from among the phosphatases (A-type esterases) while the carboxyesterase seemed to be distinct from the A-type esterases. Both were inhibitable with DDVP which, of course, inhibits esterase activity in grain. Effects of DDVP on malathion degradation and *vice-versa* in grain are discussed in section V b) 4 below.

[6] *Note added in proof:* GARDNER *et al.* (1969) have identified a malathion homologue, formed in residues on crops, as dimethyl S(1-ethoxycarbonyl-2-butoxycarbonyl) ethyl phosphorodithioate. The "unknown" observed by ROWLANDS (1968 b) as formed by malathion in the presence of wheat lipids has proved to be a similar compound.

On the "residue" side, HILL and THOMPSON (1968) and THOMPSON and HILL (1969) have surveyed existing malathion residues in cereal grains imported into the United Kingdom. MCGAUGHEY (1969) has determined the distribution in rice milling fractions and HILL and BORDER (1967) the loss of intact malathion during the malting of treated barley. KADOUM and LA HUE (1969) observed the effects of hybridization, foreign material, moisture content, and storage temperature on malathion persistence in grain sorghum. ALESSANDRINI *et al.* (1968) deduced equations for predicting the stability of malathion on stored wheat grains and ASHMAN and KEMPTON (1967) considered the nature of the base and its effect on the stability of malathion dilute dust formulations.

Cidial has been shown to have a long stability on damp grain (LONGONI and MICHIELI 1967) and ROWLANDS (1970 b) has made a preliminary assessment of the breakdown products. On grain of 18 percent moisture content, only 20 percent degradation of pure Cidial applied at ten p.p.m. had occurred after six months at 20°C. and the products identified (TLC) were Cidial acid (carboxyesterase product), desmethyl Cidial, and the phosphatase products dimethyl phosphorodithioate and phosphorothionate. These were the main derivatives with smaller amounts of monomethyl phosphorothionate. No trace of the P=O analogue (cidioxon) was found *in vivo* with whole grains but it was aerobically produced by germ lipoxidase systems *in vitro*. Degradation pathways for this carboxyester phosphorodithioate appear to be analogous with dimethoate, and possibly formothion, rather than with malathion (ROWLANDS 1967). Use of the more toxic technical Cidial on grain could give additional residues of the contaminant (S-methyl isomer?) and its phosphatase and car-

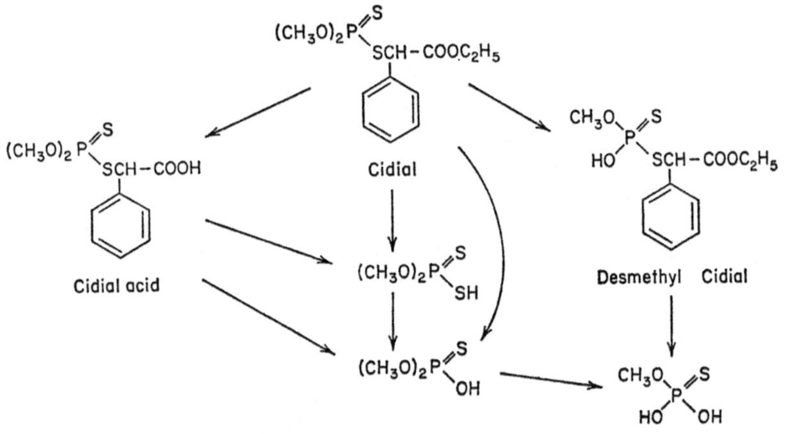

Fig. 7. Postulated metabolism of Cidial in grain

boxyesterase products. Uptake of Cidial by individual grains was rapid either from aqueous emulsion or from organic solvent. LONGONI and MICHIELI (1967) observed only 14 percent degradation of Cidial applied at ten p.p.m. to wheat grain of 13 percent moisture content during one year of storage.

V. Synergists and natural enzyme inhibitors: Their effects on metabolism in grains

Two effects that may be described as "joint action" are of concern to the analyst/biochemist/toxicologist: (a) possible prolongation of the residual life of insecticides applied to the grain and (b) potentiation (or antagonism) of the mammalian toxicity of the insecticide, or its breakdown products, to the consumer of the raw cereal or cereal products.

Compounds which may be expected to have these effects are of two basic types: natural products and additives (foreign compounds).

a) Natural products

Obviously this category embraces a huge number of substances, but while it is one of the bugbears of "pure" biochemistry that inhibitors present at cellular level are released into the general "pool" during homogenisation for *in vitro* preparations, they are usually extremely localised *in vivo*.

How and when the natural inhibitors and regulators present in grain may affect pesticide metabolism is not easily determined therefore and, in larger viewpoint, would probably be "love's labour lost". However it was thought to be worth mentioning some of those concerned, before passing on to the more important additives.

Most insecticidal application to grain obviously takes place during dormancy when cellular synthetase and amylase inhibitors, in particular, are effective. ROBERTS (1969) reviewed the factors considered to cause dormancy in cereal grains and concluded that oxidative mechanisms are frequently involved and that endogenous germination inhibitors are of great importance, there being little evidence that the covering structure of the grain imposes a limitation on oxygen uptake and germination by restricting respiration. In some cases these inhibitors increase the requirement for oxygen; in others the oxygen is needed to oxidize the inhibitor to its inactive form. Electron acceptors in particular (e.g., nitrate, nitrite, oxygen) decrease or break dormancy. During early germination the pentose phosphate pathways seem to be important and some dormancy-producing agents may inhibit this pathway; conversely some dormancy-breaking agents may stimulate germination by increasing stages of this pathway. Inhibition of cytochrome oxidase or provision of alternative electron acceptors, which can be used in the conventional respiratory processes, decrease competition for available oxygen and speed up the breaking of dormancy. Terminal oxidase, Krebs cycle or glycolysis inhibitors (including many insecticide synergists), carbon monoxide, cyanide, azide, hydrogen sulphide, and hydroxylamine, if provided, all lead to breaking of dormancy. Conversely, PEERS (1958) found germination inhibitors (not affecting oats) in aqueous extracts of oat husk. He ascribed the effects to organic acids such as succinic, fumaric, acetic, and malic. MIKKELSEN (1967) reported vanillic, ferrulic, *p*-hydroxybenzoic, *p*-coumaric, and indoleacetic acids among natural, water-soluble substances present in rice hulls; all of these acids delayed germination, as does the fungistat/bacteristat/preservative propionic acid (HUITSON 1968) [see section b) 2].

Natural "regulatory agents" of grain growth and development may be divided (BRUINSMA 1966) into primary and secondary regulators.

1. Primary regulators. These include compounds (hormones) such as kinins, auxins, and gibberellins that act via genetically controlled systems of enzyme production, by influencing genetic regulation of metabolic activity and also by interfering with enzymic activity *per se*. ROWSELL *et al.* (1966) studied the stimulation of wheat endosperm hydrolases by gibberellic acid, secreted in the embryo early in germination, and its dependence on oxygen. Similarly MACLEOD and PALMER (1967) determined the effects of gibberellins from barley embryos on the synthesis of α-amylase in the aleurone layers of barley.

2. Secondary regulators. These act chiefly upon the synthesis and formation of primary regulators but still have some direct anti-enzymic activity. Such compounds include dormin (abscissin 11), which probably competes directly with gibberellin action rather than preventing its synthesis, α-tocopherol (part of the vitamin E complex), and other fat-soluble vitamins, e.g., the saponins and vitamin K (phylloquinone) which stimulate or even replace the action of primary regulators. GREEN (1958), in suggesting an anti-oxidant role for α-tocopherol in addition to its role in cellular growth, thought that as the γ- and δ-tocopherols (the most potent anti-oxidants *in vitro*) are found mainly in highly unsaturated oils, they may protect the stored fats from oxidation. The vitamin E complex (α-, β-, γ-tocopherols) functions as a lipoxidase inhibitor *in vivo* (DRAPER 1968, WHITTING 1968). ROWLANDS (1968 a, 1970 b) showed that α-tocopherol had no effect *in vitro* on the anaerobic dechlorination of DDT to DDD but that it inhibited aerobic lipoxidase action *in vitro* and with it the coupled oxidation of certain organophosphorothioates and pyrethroids; HALL and LAIDMAN (1968 a and b) and WHITTING (1968) determined the tocopherol content of dormant wheat grains and studied the pattern and control of tocopherol metabolism in the germinating grain.

Among "proteinaceous" enzyme inhibitors are the trypsin (proteolytic, papain-type) inhibitors (LEARMOUTH and WOOD 1960, SHYAMALA *et al.* 1961) while purothionin, a low-molecular weight protein from wheat flour, is apparently derived from a lipoprotein. It has bactericidal and fungicial properties and activates wheat proteinase. FISHER *et al.* (1968) fractionated purothionin into two distinct globulin-type products (α and β). REDMAN and FISHER (1969) obtained two similar compounds (hordothionins) from barley flour; none has been found in rye, oats, or maize. ROWLANDS (1966 a) found that crude purothionin inhibited the iron-porphyrin fat peroxidizing systems of wheat germ *in vitro* under both aerobic and anaerobic conditions. It could thus affect the whole range of pesticide oxidations [section III d) 1] and also the dehalogenation reactions [section III d) 4].

Phytic acid from wheat bran inhibits α-amylase (CAWLEY and MITCHELL 1968) and the myriad naturally-occurring phenolic compounds with inhibitory action, such as salicylic, cinnammic, coumaric, and chlorogenic acids should also be mentioned. In particular, DANIELS and MARTIN (1968) have isolated 24 phenolic anti-oxidants from oats, some of which are esters of caffeic or ferulic acids with long-chain mono-alcohols or ω-hydroxy acids. DRAPER (1968) discusses the antioxidant activity of such water-soluble compounds as ascorbic acid, sulphur amino acids, and selenium compounds.

Inhibitors concerned in cereal grain or plant host-resistance mechanisms to insect predators may also be relevant. KLUN and ROBINSON (1969) reported on the concentrations of 1,4-benzoxazinones in maize at various stages of development and their role in resistance to the European corn-borer. APPLEBAUM and KONIJN (1966) described a highly-specific *Tribolium* larval protease inhibitor present in wheat which possibly constitutes a partially successful "defence mechanism" of wheat to this insect. DYTE (1967) suggested that where insect feeders have developed tolerance to such defence agents (natural inhibitors), it should be possible to interfere with any insect detoxication mechanism by use of synergists, such non-toxic compounds acting as a new type of chemical control agent. He suggested further (DYTE 1969) that some secondary plant substances may have evolved as synergists to enhance the action of other naturally-occurring protective substances. Over 300 compounds containing the methylenedioxypheny group [essential to the action of natural and synthetic benzodioxole synergists (WILKINSON 1965)] are known to occur in plants (NEWMAN 1962). For example, dihydrosafrole, isosafrole, myristicin, and safole (naturally-occurring) all synergise insecticides. REIFER *et al.* (1968) have compiled a useful list of inhibitory substances in plants.

b) Additives

The effects of compounds foreign to the grain environment are easier to determine, or reasonably predict, than are those of the localised natural products. There are numerous "additives" that can be distributed in greater or lesser amounts throughout a treated grain bulk, such as synergists, antimicrobials, and existing pre- and post-harvest residues of other insecticides and fumigants.

1. Synergists. Insecticide synergists are compounds which are themselves nontoxic at the dosages employed but which are able to enhance the toxicity of the insecticide with which they are applied. At the present time only the methylenedioxyphenyl compounds (with pyrethroids and anti-oxidants) are applied to stored grain. However, spread of resistance to insecticides among stored product insects, coupled with the paucity of research on alternative control measures, might necessitate the use of many more synergistic formulations against stored grain pests. This section will, therefore, be covered in more detail than their current usage would seem to warrant.

Since, in general, synergists act by inhibiting insect detoxication mechanisms, it is probable that they will have similar effects on the grain physiological processes. A number of behavioural and "physical" mechanisms of synergism have also been described. METCALF (1967) reviewed the mode of action of insecticide synergists and WILKINSON (1968) reviewed their role in overcoming or elucidating insect resistance problems.

Synergists that decrease the ability of an organism to metabolise (usually detoxify) an insecticide are of several types (DYTE and ROWLANDS 1967): (a) inhibitors of microsomal oxidases which metabolise a variety of foreign compounds in insects, mammals, and plants and which may affect the metabolism of a given insecticide in several different ways; (b) analogue synergists, structurally related to the insecticides they synergise. Thus the phosphate synergists can be regarded as analogues of the organophosphorus insecticides, much as DMC is used with DDT. Analogue synergists can be specific in that they are effective with only one type of insecticide and may affect only one detoxication pathway. By contrast the various pyrethroid synergists are structurally unrelated to the insecticides they affect; (c) synergism of organophosphates by non-insecticidal carbamates and the synergism of carbamates by non-insecticidal phosphates [reported by PLAPP and VALEGA (1967)] represents a third type of synergism. These synergists are not analogues of the insecticides but analogues of other insecticial compounds which inhibit the same target enzyme. Some of them, such as S,S,S-tributyl phosphorotrithioate and butyl *O*-methyl carbanilate, are also effective analogue synergists; (d) DYTE and ROWLANDS (1967) suggested that there might be another class of synergist. SKF 525A is a non-analogue synergist of type (a) in that it inhibits the oxidative metabolism of a variety of drugs and insecticidal compounds. However, it also inhibits detoxification by insect phosphatases which are unaffected by the majority of type (a) synergists. Since the organophosphorus insecticides must have an affinity with both the hydrolases which detoxify them and with the target cholinesterase, which they inhibit, it seems probable that alternative substrates for the detoxifying enzymes might not only be phosphate analogues, and analogues of other cholinesterase inhibitors such as carbamates, but also analogues of acetylcholine itself. SKF 525A might well be regarded as such a substance:

$CH_3COO(CH_2)_2N(CH_3)_3$	$C_3H_7(Ph)_2COO(CH_2)_2N(CH_2CH_3)_2+$
acetylcholine	SKF 525A

The methylenedioxyphenyl (1,3-benzodioxole), or MDP, synergists are the most important. Among them are piperonyl butoxide, sesamex, sulfoxide, and tropital which are used commercially as pyrethroid synergists. Piperonyl butoxide and sulfoxide were among compounds tested for tumorigenicity, against mice, whose effects were sufficiently dubious to require further evaluation and testing (INNES *et al.* 1969).

It is supposed that efficient MDP compounds fit well at the active sites of the insecticide detoxifying enzymes, but are only slowly degraded, the fit involving both the planar benzodioxole ring and the side chain (ESSAC and CASIDA 1969, WILKINSON and HICKS 1969). BOWERS (1968) has shown that many MDP compounds, natural and synthetic, have high insect juvenile hormone activity and species specificity. MDP compounds are metabolised by the same oxidative, microsomal-NADPH hydroxylating system that metabolises (detoxifies) certain carbamate and pyrethroid insecticides (ESAAC and CASIDA 1969). In reducing the amount of insecticide metabolism, they act therefore as competitive inhibitors of the microsomal mixed-function oxidases responsible for detoxication, possibly at some specific site on the electron transport chain; the terminal oxidase may be the microsomal cytochrome P450, for instance. ESAAC and CASIDA (1969) have shown that metabolism of MDP compounds in living houseflies is by one or more of the following systems: oxidative attack at the methylene carbon resulting in scission of the MDP moiety, oxidative attack at the aliphatic side chain, and conjugation of aromatic acid residues (resulting from such side chain oxidation) as β-D-glucosides or as amino acid derivatives. Side chain oxidation occurs at the alcohol group of piperonyl alcohol, the aldehyde group of piperonal, the acetal group of tropital, each of the three aliphatic ether groups of piperonyl butoxide, the unsaturated side chain of safrole or isosafrole, and the sulfinyl group of sulfoxides A and B. Degradation of the MDP moiety seems to proceed according to the following scheme:

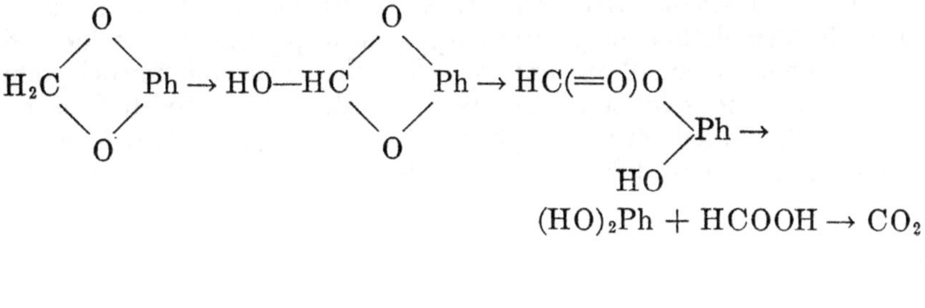

Of the MDP compounds applied directly to grain in practice, only piperonyl butoxide has received extensive use and then only in formulation with pyrethroids. Most commercial pyrethrin formulations contain this synergist and sometimes other anti-oxidant stabilizers which increase shelf-storage life of the constituents and decrease photolysis. BROOKE (1967) has shown that MDP compounds themselves have little or no effect on this aspect of pyrethroid decomposition.

MDP compounds penetrate readily into grain upon application (STRONG *et al.* 1961, ROWLANDS 1970 b) and are slowly degraded (BLINN *et al.* 1959). ROWLANDS (1966 b) showed that piperonyl butoxide inhibited phenoloxidases in the seedcoats of freshly-harvested wheat grain and prevented the coupled oxidation of malathion and bromophos to their P=O analogues in *vitro;* the enzymes slowly recovered, *in vivo,* presumably as the MDP compound was itself degraded. WILKINSON (1965) has shown that MDP compounds inhibit mushroom tyrosinase *in vitro.* ROWLANDS (1970 b) later found that piperonyl butoxide, sesamex, and tropital inhibited the aerobic oxidation of phosphorothioates by wheat grains *in vivo* and by wheat germ lipoxidase *in vitro,* also the *in vitro* oxidation of pyrethroid-I type compounds. The MDP compounds did not affect anaerobic reactions of the same system, which dechlorinated lindane and DDT, nor did they have any effect on the phosphatase/esterase hydrolysis of organophosphorus and pyrethroid compounds or upon the phosphatase demethylation of dimethyl phosphorothionates. Thus it would appear that stable MDP compounds could be expected to prevent, or at least delay, oxidative metabolism of phosphorothioates and possibly pyrethroids in grain but have no effect on their hydrolytic degradation. Since the germ oxidative mechanisms affecting phosphorothioate toxication only seem to be effective on a long-term basis ($>$ eight months storage) this is unlikely to be of practical significance; considerable degradation of both the MDP compound and the phosphorothioate would be expected after such an interval. Initial oxidation by grain phenolases is of very little importance and pyrethroids applied to grain *in vivo* apparently undergo hydrolytic degradation only. ROWLANDS (1970 b) observed that piperonyl butoxide was degraded by grain enzymes *in vivo,* apparently to a compound similar to piperonyl alcohol and by germ lipoxidase *in vitro* to hydroxyaromatic compounds, formate, and CO_2.

The antioxidants BHT and BHA are frequently added as stabilizers in formulations of pyrethrins/pyrethrins-piperonyl butoxide and, in consequence, small amounts may find their way into treated grain. Their chief use is for stabilizing fats and oils used in foodstuffs, sometimes with added citrate or phosphate, and in the United Kingdom residues of 100 to 1,000 p.p.m. are permitted in certain foods as a result of such usage. In many countries (e.g., Canada, The Netherlands), only residues up to 100 p.p.m. are acceptable and there is growing concern about the long-term effects of BHT. No tolerance has been allowed in Australia, although residues of up to 100 p.p.m. BHA are permitted. The UN/WHO Expert Committee on Food Additives (1965) has decided that further work on the long-term toxicity of BHT to mammals is desirable. ROWLANDS (1966 b) found that BHT inhibited the transient phosphorothionate oxidase (phenolase) of wheat grain seedcoats and that the inhibitor did not apparently degrade during storage of treated wheat, nor did the oxidases recover; metabolism in mammals seems to proceed via oxidation of the butyl or methyl groups. At high dosages germination of the grains was impaired. BHT had little or no effect on hydrolases at doses that completely inhibited oxidase activity.

A number of compounds related to SKF 525A increase the effects of drugs in mammals and synergise pyrethroids against insects (BATES *et al.* 1965). DYTE and ROWLANDS (1967) showed that SKF 525A inhibited oxidation of malathion to malaoxon by flour beetles *in vivo* and that it also inhibited their degradation of malathion by phosphatases but not by carboxyesterases. The former effect probably accounted for the overall antagonism observed with SKF analogues against malathion and, as anticipated, SKF 525A synergised the actual toxicant, malaoxon, *per se* (*i.e.*, it inhibited phosphatase breakdown of malaoxon) whereas other MDP compounds that do not inhibit phosphatases, did not. In addition SKF 525A reduced the "oxon" and desmethyl derivative levels in *Tribolium castaneum* treated with bromophos or Sumithion. This inhibition of phosphatase degradation is potentially synergistic and, since there is no reason why the structural requirements for the two effects (antioxidant and antiphosphatase) should be similar, DYTE *et al.* (1968 and 1969) screened some 30 analogues of SKF 525A to test their effect on the potency of malathion against the flour beetle. Unfortunately the structural requirements to select the phosphatase inhibition could not be elucidated and the only conclusion that could be drawn was that the range of analogues antagonising malathion against *T. castaneum* was greater than that synergising pyrethrins against the lesser mealworm beetle. Many of the synthetic hydrazine plant-growth retardants (auxin inhibitors) such as B-9 and C-O11 (REED 1965) can be considered as analogues of SKF 525A.

When applied to wheat grains SKF 525A inhibited germ phosphorothioate oxidation and phosphatase hydrolysis (ROWLANDS 1970 b); seedcoats of freshly-harvested wheat steeped in solutions of SKF 525A lost their transient phenoloxidase-phosphorothioate oxidase activity but slowly recovered due to degradation of the inhibitor. Wheat germ lipoxidase preparations were also inhibited *in vitro* and SKF

525A was rapidly degraded hydrolytically both *in vitro* and *in vivo* to substituted phenylacetic acids and ethylamino alcohols (ROWLANDS 1970 b).

Some noninsecticidal organophosphate compounds inhibit phosphatase and esterase breakdown of their insecticidal analogues by mammals and insects and, although the efficiency of such compounds has been tested as synergists for malathion against resistant houseflies of different strains (PLAPP and TONG 1966) and against resistant and susceptible flour beetles (DYTE and ROWLANDS 1967 and 1968), little is known of the specificity of these compounds for different detoxication pathways. Moreover, the application to insects of two phosphorus compounds together may result in decreased penetration of at least one of them and even consequent reduction in toxicity of the active compound when a noninsecticidal compound is used with an organophosphate pesticide (ABDALLAH 1963, DYTE *et al.* 1966). Similar effects have been observed with sesamex and some organophosphate insecticides (MENZER and CASIDA 1965). Triphenyl phosphate (TPP) overcame malathion-specific resistance in flour beetles (DYTE and ROWLANDS 1968) by inhibiting carboxyesterase but not phosphatase breakdown of malathion in both susceptible and resistant strains. However, since TPP synergises malathion in mice (CASIDA 1961), and probably does so by inhibiting mammalian carboxyesterase, it is of no use for practical control measures against malathion-specific resistance of the enhanced carboxyesterase type in stored food pests.

TPP prevented hydrolysis of malathion or malaoxon, at carboxyester and thiolo-carbon linkages but not at the thiophosphate bond, by wheat germ preparations *in vitro* and it prolonged the effective life of malathion applied to damp grains, as did tri(*ortho*)cresyl phosphate (TOCP) (ROWLANDS 1967). The toxicological effects of both compounds in mammals would preclude their use for this purpose in practice. TPP produced only slight phosphorylation of wheat protein in contrast to that with phosphates such as DDVP, dibrom, etc.

Other synergists which might be mentioned briefly in passing are uncouplers of oxidative phosphorylation (WEINBACH and GARBUS 1969) such as oligomycin (KENEFICK and HANSON 1966), Thanite, and octachlorodipropyl ether (FISHBEIN *et al.* 1968, DYTE and ROWLANDS 1967), aryl propynyl ethers or "propargyl" compounds (SACHER *et al.* 1968, BARNES and FELLIG 1969, DYTE and ROWLANDS 1970 a), thiocyanates (BAKRY 1961), and aryl diethyl sulphides (MANDEL'BAUM *et al.* 1968).

2. Anti-microbials. Cereals of high moisture content constitute better animal-feed ingredients than dry grains and, of course, a bulkier product to sell by weight. However, the risks of spoilage in storing high moisture grain under normal conditions are obvious (ROWLANDS 1967): heating, mould growth, and sprouting. To facilitate high moisture storage the use of certain preservative sprays, chiefly simple aliphatic fatty acids such as formic, acetic, and propionic, have recently been developed. HUITSON (1968) has reported the use of propionic acid spray as a preservative for high moisture grains. It had, apparently, fungistatic and bacteristatic properties, suppressed germination (hence could not be used on "malting" barley), and appeared to inhibit mite development. There was, however, no evidence that it would prevent mould spreading from untreated areas. The use of such a propionic acid spray on grain has been approved in the United Kingdom for animal feed provided that dosage rates of 0.3 to 1.5 percent, according to the moisture content, on the weight of grain (*i.e.*, 334 to 1,563 p.p.m.) are not exceeded (1 percent = two gal./ton, approximately). At present it may only be used on grain intended for animal feed but there is always a possibility that it may be extended to grain for human consumption in the future; a rate of ¾ lb. of acid/ton of grain has been suggested. Tainting problems arise, however, and might lead flour millers to reject such grain. Propionic acid and its sodium and calcium salts are permitted in the United Kingdom for use as preservatives in bread flour and confectionery, at levels up to 3,000 and 1,000 p.p.m., respectively, and SCHROEDER (1964) has reported on the use of sodium propionate at 5,000 p.p.m. to preserve rough rice from fungal development. Technical propionic acid (*i.e.*, the commercial formulation), which is rapidly absorbed by grain, contains approximately 2.5 percent of formic and acetic acids also and necessitates the use of special corrosion-proof spraying equipment.

WILKIN and THOMAS (1970) have shown that propionic acid applied at one percent to stored barley has slight initial insecticidal action *per se* but that it confers no protection against infestation by insects.

The direct application of concentrated aliphatic acids to grain at high levels, such as those recommended, is bound to bring about many physical and environmental changes. At the 1.5 percent level of propionic acid application (approximately three gal./ton), the overall pH of wheat grains decreased from an average of six to 4.5. Grain pH is an obviously influential factor on enzyme catabolism and, as already noted, simple aliphatic acids are among pronounced germination and metabolism inhibitors (PEERS 1958), although being normal "intermediates" in metabolic cycles. Formic acid has been shown to be a potent enzyme inhibitor and to interfere with catalase in particular (see UN/FAO 1965 for references).

The effect of such an acid treatment on any insecticide residues present in grain needs very careful consideration and study—the more so because of the possible complex of reactions involved. Mere decrease in overall grain pH alone can considerably enhance nonenzymic degradation of many insecticide residues [see ROWLANDS (1967) for discussion of pH effects]; high fatty-acid content, for example, enhances nonenzymic hydrolysis of malathion in rice brans and other cereal products. DYTE *et al.* (1970) found that both formic and acetic acids topically applied slightly synergised sumithion against adult flour beetles but that propionic acid slightly antagonised its effect. Formic acid did not affect the toxicity of ronnel, dimethoate, Abate, imidan, or Methyl Trithion but antagonised the carbamate C-8353. All three aliphatic acids inhibited the production of sumioxon and desmethyl sumithion from sumithion in flour beetles but did not appear to affect attack on the P-O-phenyl bond. These effects are directly comparable to those of the MDP compounds and, of course, ESAAC and CASIDA (1969) have suggested that oxidation of the methylene group to formic acid and other derivatives plays a role in the metabolism and possible mode of action of the MDP synergists. The parallel effects are perhaps not surprising, therefore. Formic acid and formaldehyde were unstable in their housefly abdomen-NADPH system and less than 50 percent of ^{14}C-sodium formate injected into flies was given off as CO_2. This agrees with results obtained for formate metabolism in mammals (see UN/WHO 1965) where considerable species difference in levels excreted and expired as CO_2 are to be observed. The results with insects suggest that the toxicological effects of mixtures of aliphatic acids and insecticides as residues merit further study, especially on mammalian systems.

HORLER *et al.* (1968) showed that the potent fungistat 8-hydroxyquinoline had no effect on the breakdown of sumithion, malathion, or bromophos in barley of 12 percent moisture content during three weeks storage. They also found (unpublished) very slight retarding effects with this fungistat in barley of 19 percent moisture content; these "delaying" effects were marginally significant up to 16 days storage but not thereafter.

Zymocidal food preservatives, such as sorbic acid, potassium sorbate, and benzoic acid have been shown to have insecticidal action against flour beetles by suppressing vitamin-synthesising microorganisms within the insects which then died from vitamin deficiency (MILNE 1964). However, they are probably of insufficient activity to warrant consideration of their use in practice.

Residues of trihalophenols, from ronnel, bromophos, iodophos, etc. may possess antifungal activity (see earlier); dichlorobromophenol from bromophos prevented further *in vitro* breakdown of bromophos by *Alternaria tenuis* but did not affect the exogenous metabolism. Other effects of the halophenols will be considered below, as will such volatile antimicrobials as ethylene and propylene oxides.

3. Existing fumigant residues. Grain that has been disinfested by fumigation frequently needs subsequent protection with a contact insecticide during storage. Conversely, if insecticide treated grain that has become infested has to be sold, it may be fumigated to kill the infestation. Residues of fumigants and their degradation products may therefore be present with those of contact insecticides in stored grain and can effect the persistence and toxicology of the insecticides residues. A wealth of data on fumigant residues encountered in practice and in experimental use have been usefully surveyed by LINDGREN *et al.* (1968) and, unless new matter is introduced, references to original authors are omitted here since their contributions have been fully detailed in that review paper.

Methyl bromide (MB) is one of the most important fumigants for disinfestation of stored grains and THOMPSON (1968) has reviewed its use and properties. MB decomposes within the grain to give inorganic bromide and methylated derivatives of —OH, —SH and —NH groups. As might be anticipated, the proteins account for 80 percent of this reaction and the majority of methylated derivatives occur in the water-soluble (glutenin) fraction. *N*-methyl derivatives comprise approximately 50 percent of the total reaction with the glutenins and highly unstable dimethyl sulphonium derivatives about 30 percent. The major combination products with wheat are not detrimental to the food value and are nontoxic to humans.

ROWLANDS (1969) found fumigation with MB, that left residues of 20 p.p.m. inorganic bromide after aeration, prevented desmethylation of sumithion applied one week later to the grain, presumably by methylating receptor sites or the desalkylation (phosphatase) system (Fig. 8). This level of methylation did not greatly affect the P-O-aryl phosphatase hydrolysis, however, nor did more intense fumigation that left residues of up to 120 p.p.m. of bromide. He also found (unpublished) that levels $>$ 120 p.p.m. of inorganic bromide decreased the reductive dechlorination of DDT under anerobic conditions. This could be due to methylation of active sites of enzymic protein, to "swamping" of receptors of liberated halide in the grain or, most likely, by successful competition for the iron-porphyrin compounds, since CASTRO (1964) has shown that dilute solutions of reduced iron-porphyrins are oxidized by certain alkyl halides at room temperature to form haemins (Fe^{3+}-halides), analogous with the reaction proposed for DDT [section III d) 4]. More active alkylating agents such as methyl iodide, dimethyl sulphate, or ethylene oxide decreased more of the general enzymic activity of the grain than did MB and completely blocked the desmethylation of sumithion. Prior fumigation with MB had no effect on the residual life of malathion, which does not appear to be desmethylated in grain. HOLLINGWORTH (1969) suggested that the probable route of *O*-dealkylation of dimethyl (thio-)phosphates in mammals is by S-alkylation of glutathione. Similarly, other workers showed that mammalian tissue preparations produced S-alkyl glutathiones with alkylating agents like methyl iodide, ethylene halohydrins, and ethylene dibromide (see HOLLINGWORTH 1969 for references). The dealkylation of methyl paraoxon by mammalian tissue preparations was strongly

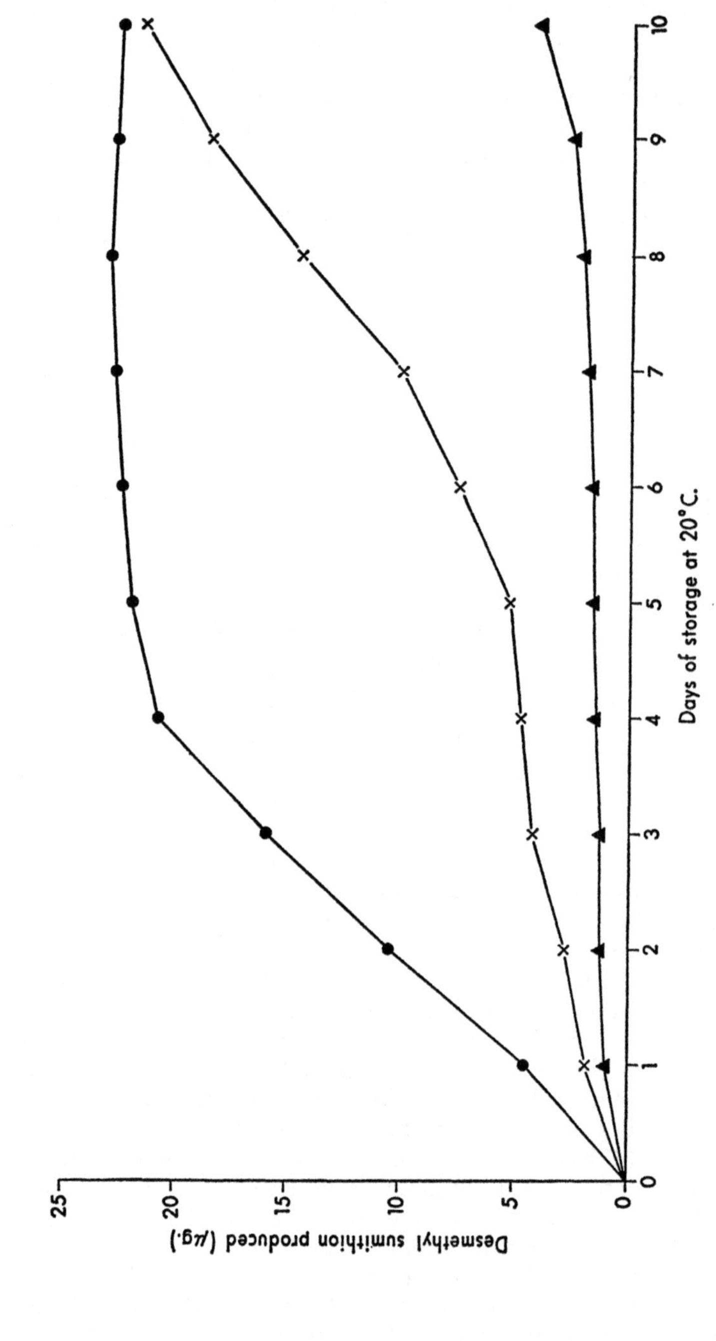

Fig. 8. Effects of methyl bromide residues on sumithion desmethylation in wheat of 18 percent moisture content: ●——● untreated wheat, ×——× wheat containing 20 p.p.m. of inorganic bromide, and ▲——▲ wheat containing > 120 p.p.m. of inorganic bromide

inhibited by methyl iodide which, as anticipated, synergised sumithion against mice.

Ethylene dibromide (EDB) is frequently used with other fumigants such as ethylene dichloride, carbon tetrachloride, and MB for the treatment of cereal grains. When used alone, little "biochemical" reaction occurs between EDB and wheat at room temperature, despite its rapid sorption.[7] Such products as have been obtained at room temperature suggest methylation of the proteins, but to a much lesser extent than with MB. Any effects upon insecticide metabolism observed with residues from MB treatment could therefore be anticipated with EDB alone, but to a lesser degree. Mixtures of MB and EDB have been used on a variety of cereals. Almost complete separation of the components of such a mixture occurred during the penetration of a few inches of flour. In many parts of a bulk therefore, MB is likely to be the sole fumigation (alkylating) agent and residues will be consistent with that level of MB treatment.

Ethylene dichloride (EDC) is a chemically inactive fumigant that does not decompose biochemically in treated wheat; mixtures of EDB, EDC, and carbon tetrachloride have been used. The biochemical fate of carbon tetrachloride in wheat has not been studied, nor those of the related compounds with which it may be mixed, such as carbon disulphide and chloroform. However, as with compounds such as hydrogen cyanide, acrylonitrile, methalyl chloride, chloropicrin, sulphuryl fluoride, and methanesulphenyl chloride, its limited use in practice obviates any detailed consideration in this context.

Ethylene oxide (EO) is highly inflammable and its occasional experimental use on stored cereals is almost invariably in combination with CO_2 or nonflammable halogenated hydrocarbons. It is also used in some circumstances as a sterilant, to destroy microbial populations, as is propylene oxide. EO is quite a potent alkylating agent, more so than MB, and ethylation within wheat grain results in interference with desalkylation of sumithion and a decrease in levels of enzymic activity generally (Rowlands, unpublished data). Among reactions affected was the anaerobic reductive dechlorination of DDT, but this particular inhibition cannot positively be ascribed to akylation of the enzyme systems responsible, since EO is known to react with naturally-occurring chlorides to form the more toxic ethylene chlorohydrin (2-chloroethanol, ECH) in foodstuffs. Heuser and Scudamore (1969) have reported on the analogous and preferential formation of ethylene bromohydrin (EBH) in wheat and wheat flour during fumigation with EO, when the required bromine was derived either from naturally-occurring inorganic bromide or from bromide produced by prior fumigation with MB or EDB. They found that the EBH produced

---

[7] On heating some decomposition to ethylene glycol, which may react with —SCH_3 of methionine, occurs in wheat.

in flour persisted in conditions of limited air movement but broke down more rapidly than did ECH under the same storage conditions at 25°C. This might indicate partial breakdown of the EBH due to continuous production of alkali by the slower reaction of adsorbed EO with available chloride. When treated at sterilizing levels, the flour acidity changed from pH 5.6 to 6.2 after six hours exposure, reaching a maximum of 6.5 after some days. While the ECH content continued to increase during seven days' storage, the EBH reached a maximum at 24 to 48 hours, although considerable free EO was still present. There is little published information on the mammalian toxicology of ethylene halohydrins but they seem to act as alkylating agents also and to form S-ethyl glutathiones in mammalian preparations, analogous with those of dimethyl (thio-)phosphates, methyl iodide, EDB, and EO. Similar hydrin derivatives, 1-halo-2-propanols, are formed, but to a lesser extent, by propylene oxide in cereals and flours.

Phosphine is usually generated *in situ* by exposure to moisture of tablets or pellets containing aluminum phosphide and ammonium carbamate. Residues remaining can be expected to include phosphine, phosphorylated proteins or amino acids, unreacted aluminum phosphide formulation, and aluminum hydroxide from reaction of water and aluminum phosphide. Washing the grain removes most of these products. The phosphorylated products, phosphites, may well have similar effects upon residual insecticides to those found after DDVP treatment of grain [Rohrlich and Meuser (1969), also see below]. Berck (1968) inferred binding of phosphine in some form to protein constituents, sorption losses being particularly high in the proteinaceous fractions of cereals.

4. Existing and concurrent insecticide residues. Insecticides used as seed dressings or applied to the growing cereal crops will be metabolised and varying amounts of the intact pesticide or metabolic products will translocate to the maturing grains. For example, a detailed study by Tomizawa and Sato (1962) elucidated malathion and methyl parathion metabolism, translocation, and final concentration of metabolic products in the unripe grains of rice plants. Laws (1965) has reviewed a good deal of information on residues persisting at time of harvest. If such residues are present they will undoubtedly affect the initial degradation of contact insecticides subsequently applied in store. The following insecticides are among those frequently used for soil or cereal crop treatment and, even after the recommended aging period, some residues may remain in the harvested grain: DDT, demeton, demeton-S-methyl sulfoxide, diazinon, dimethoate, ethion, fenthion, lindane, malathion, methoxychlor, methyl parathion, parathion, phosphamidon, Sevin, sumithion, trichlorfon, and Trithion. In addition, methyl bromide and ethylene dibromide are sometimes used for soil treatment before planting. Mention should also be made perhaps of the possibility of interaction with translocated herbicides. For example, BOWLING and FLINCHUM (1968) examined the interaction of the herbicide propanil with organophosphorus, organochlorine, and carbamate insecticides applied as seed dressings to rice.

Table XII summarises the known retarding effects of insecticide breakdown products on grain enzymes. As shown in this table, some

Table XII. *Demonstrated effects of some insecticides and their breakdown products or grain enzymes*[a]

Compound	Enzyme(s) affected	Nature of retarding effect[c]			
		Inhibition[b]	Competes	*In vitro*	*In vivo*
Parent phosphates	Lipoxidase	✓	—	✓	—
or	(Carboxy)esterase	✓	—	✓	✓
phosphorothioate	Esterase (general)	✓	—	✓	✓
"oxons"	Decarboxylase[d]	✓	—	✓	—
Desmethyl bromophos	Alkylphosphatase	✓	—	✓	✓
	Arylphosphatase	—	✓	✓	✓
Desmethyl Ronnel	Alkylphosphatase	✓	—	✓	✓
	Arylphosphatase	—	✓	✓	✓
Desmethyl Sumithion	Alkylphosphatase	✓	—	—	✓
	Arylphosphatase	—	✓	—	✓
Dichlorobromophenol	Arylphosphatase	✓	—	—	✓
Trichlorophenol	Lipoxidase	✓	—	✓	—
1,2,4-Trichlorobenzene	Lipoxidase	✓	—	✓	—
(TCB)	Oxidases (general)	✓	—	✓	—

[a] This table includes only compounds that have been specifically tested for anti-enzyme activity and does not cover obvious "competitors". For example, phosphates and phosphorothioates will compete for phosphatase sites.

[b] By phosphorylation, methylation, etc.

[c] — means reaction not investigated.

[d] The germ enzyme decarboxylating malathion carboxyesterase products.

of the normal metabolic products of insecticides applied to grain may retard further breakdown. These compounds are usually fairly stable and accumulate, the exceptions being the phosphorothioate "oxons" whose inhibitory roles have only been established *in vitro* as far as grain enzymes are concerned. As already mentioned, certain trihalophenols (and possibly other phenolic residues) from aryl phosphorothionates prevent further degradation of their parent compounds in fungal cultures. This also seems to occur to a limited extent in grain by inhibition of phosphatases; they also inhibit lipoxidase *in vitro*. It is not yet understood how much of the apparent inhibitory activity of certain desmethyl aryl phosphorothionates may be due to these phenolic constituents. Certain di- and trihalophenols uncouple oxidative phosphorylation (LOOMIS 1949) and could therefore be expected to present some mammalian toxicological hazard as well as affecting the grain metabolic cycles [2,4,6-trichlorophenol requires further evaluation of its tumorigenic effect against mice (INNES *et al.* 1969)]. Desmethyl bromophos has been shown to retard further breakdown of bromophos in grain. ROWLANDS (1966 a) suggested the following mechanism: attack on bromophos by "alkyl" phosphatase (desmethylase) to produce desmethyl bromophos which accumulates to a level that temporarily, or perhaps reversibly, inhibits the phosphatase isozyme concerned. During this inhibition period, that part of desmethyl bromophos remaining unbound competes successfully with bromophos for hydrolysis catalysed by "aryl" phosphatase (attacks P-O-aryl bonds) and is degraded to monomethyl phosphorothionate and dichlorobromophenol. Desmethyl derivatives of Ronnel and sumithion act similarly but to a lesser degree.

In grain stored for less than eight months after harvest, phosphorothioate "oxons" are not likely to occur to an extent that might affect insecticide persistence; it should be remembered that these compounds can, however, inhibit a range of grain enzymes. Malaoxon, for example, *in vitro* will inhibit aerobic lipoxidase, lipase, amylase, esterases generally and malathion-carboxyesterase in particular, but not the alkyl phosphatases (desmethylase and/or the enzymes catalysing attack on the P-S-C bonds in malathion and dimethoate) or aryl phosphatases (ROWLANDS 1967).

When similar classes of compound are present in grain they can be expected to compete for enzyme sites. Thus, all other considerations being equal, malathion would compete with other phosphates and thiophosphates for phosphatases, with other phosphorothioates for the P=S→P=O oxidase and possibly with acethion, dimethoate, Cidial, or formothion for the carboxyesterase/amidases. Similarly, DDT might compete with methoxychlor or lindane for the germ de(hydro)-chlorinase.[8] However, there are other interactions to consider, particularly when grain has received a postharvest treatment that has failed or degraded and which, for various reasons, is retreated with the same, or different, insecticide. Under these conditions there may be persisting residues whose interactions are likely to be complex. As an example of this situation, the interaction of malathion and DDVP residues in grain can be cited.

If grain treated with malathion became infested and rapid disinfestation was necessary (*i.e.*, for immediate sale) fumigation might

[8] These examples are, of course, all cases of the "analogue synergism/antagonism" discussed earlier.

be one solution. It has been suggested that DDVP might possibly be used to achieve this and to obtain some (slight) additional protectant effect. If the failure of malathion to control was due to resistance, for example, there could still be high residues of malathion present. None of the malathion metabolites with the exception of malaoxon are known to have any inhibitor effect on grain enzymes, so that the only potential "interactant" would be malathion *per se.* Rowlands (1970 a) found that in wheat grain of 18 percent moisture content containing a residue of 8.5 p.p.m. of malathion, treated two weeks earlier at ten p.p.m., and then treated at ten p.p.m. with DDVP, the degradation of DDVP was slightly retarded (Fig. 9). This is

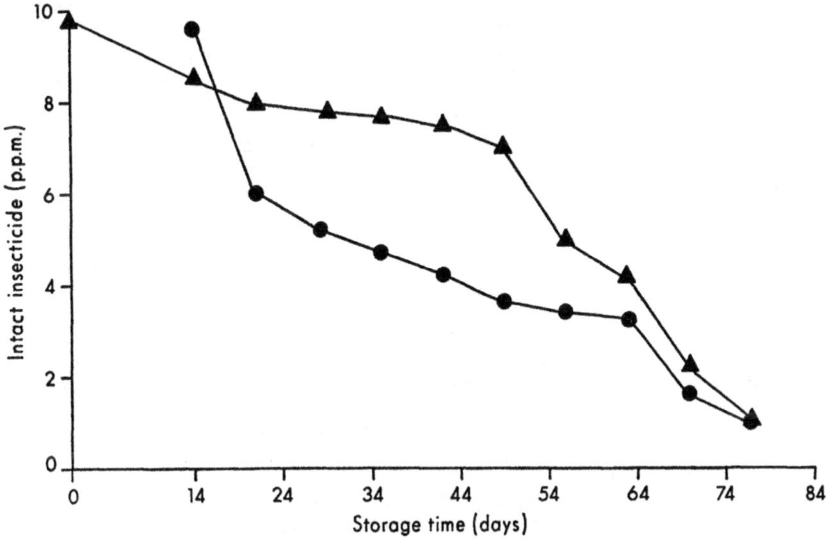

Fig. 9. Breakdown rate of DDVP applied to wheat pretreated with malathion at ten p.p.m., two weeks previously: ▲——▲ malathion, and ●——● DDVP

thought to represent competition from malathion for the alkyl phosphatases, since malathion does not inhibit any grain enzymes. It is certain also that DDVP had some reciprocal retarding effect on the subsequent rate of the malathion degradation (cf. Fig. 10).

Conversely, grain rapidly disinfested with DDVP might be treated with another contact insecticide, or even with a joint formulation initially, to give a more persistent protective effect against reinfestation. Rowlands (1970 a) examined the effects on malathion and bromophos breakdown, in wheat of 18 percent moisture stored at 20°C., of simultaneous treatment at ten p.p.m. with DDVP. Figure 10 shows that ten p.p.m. DDVP retarded the degradation of both

Table XIII. *Chemical names of pesticides and related compounds mentioned in text*

Pesticide	Chemical designation
Abate	*O,O,O′,O′*-tetramethyl *O,O′*-thiodi-*p*-phenylene phosphorothioate
allethrin	allethronyl chrysanthemate
allethrolone	2-allyl-4-hydroxy-3-methyl-2-cyclopenten-1-one
BHT	2,6-dibutyl-β-cresol
bromophos	*O,O*-dimethyl *O*-(2,5-dichloro-4-bromophenyl) phosphorothioate
C-8353	*O*-(1,3-dioxolan-2yl) phenyl methylcarbamate
chlorthion	*O,O*-dimethyl *O*-(3-chloro-4-nitrophenyl) phosphorothioate
chrysanthemic acid	chrysanthemum monocarboxylic acid
Cidial	*O,O*-dimethyl *S*-ethoxycarbonyl benzyl phosphorodithioate
cinerolone	2-butenyl-4-hydroxy-3-methyl-2-cyclopenten-1-ene
DDA	bis(*p*-chlorophenyl) acetic acid
DDD	1,1-dichloro-2,2-bis(*p*-chlorophenyl) ethane
DDE	1,1-dichloro-2,2-bis(*p*-chlorophenyl) ethylene
DDT	1,1,1-trichloro-2,2-bis(*p*-chlorophenyl) ethane
DDVP	dimethyl 2,2-dichlorovinyl phosphate
demeton	*O,O*-diethyl *O*(and *S*)-2-(ethylthio)ethyl phosphorothioates
diazinon	*O,O*-diethyl *O*-(2-*iso*propyl-4-methyl-6-pyrimidinyl) phosphorothioate
dibrom	dimethyl-1,2-dibromo-2,2-dichloroethyl phosphate
dicapthon	*O,O*-dimethyl *O*-(2-chloro-4-nitrophenyl) phosphorothioate
dimethoate	*O,O*-dimethyl *S*-(*N*-methylcarbamoylmethyl) phosphorodithioate
dimethoate carboxylic acid	*O,O*-dimethyl-*S*-(carboxymethyl) phosphorodithioate
dimethrin	2,4-dimethyl-benzyl chrysanthemate
dipterex	*O,O*-dimethyl (2,2,2-trichloro-1-hydroxyethyl) phosphonate
dursban	*O,O*-diethyl *O*-(3,5,6-trichloro-2-pyridyl) phosphorothioate
ethion	*O,O,O′,O′*-tetraethyl *S,S′*-methylenebis phosphorodithioate
fenthion	*O,O*-dimethyl *O*-(4-methylthio-3-tolyl) phosphorothioate
formothion	*O,O*-dimethyl *S*-(*N*-formyl-*N*-methylcarbamoylmethyl) phosphorodithioate
Gardona	β-isomer of dimethyl (2-chloro-1-(2′,4′,5′-trichlorophenyl) vinyl phosphate
Guthion	*O,O*-dimethyl *S*-(4-*exo*-1,2,3-benzotriazio-3(4H)-yl methyl phosphorodithioate
imidan	*O,O*-dimethyl *S*-(phthalimidomethyl)phosphorodithioate
iodophos	*O,O*-dimethyl (2,5-dichloro-4-iodophenyl) phosphorothioate

Table XIII. (Continued)

Pesticide	Chemical designation
Kelthane	4,4′-dichloro-α-(trichloromethyl) benzhydrol
lindane	γ-isomer of 1,2,3,4,5,6-hexachlorocyclohexane
malaoxon	*O,O*-dimethyl *S*-(1,2-dicarbethoxy)ethyl phosphorothiolate
malathion	*O,O*-dimethyl *S*-(1,2-dicarbethoxy)ethyl phosphorodithioate
malathion dicarboxylic acid	*O,O*-dimethyl *S*-(1,2-dicarboxy)ethyl phosphorodithioate
malathion monocarboxylic acid	*O,O*-dimethyl *S*-(1-ethoxycarbonyl-2-carboxy)ethyl phosphorodithioate
methoxychlor	1,1,1-trichloro-2,2-bis(*p*-methoxyphenyl) ethane
methoxy-DDE	1,1-dichloro-2,2-bis(*p*-methoxyphenyl) ethylene
methyl parathion	*O,O*-dimethyl *O*-(*p*-nitrophenyl) phosphorothioate
mobam	benzothion-4-yl methyl carbamate
NRDC 104	5-benzyl-3-furylmethyl (+) *cis-trans*-chrysanthemate
NRDC 108	5-benzyl-3 furylmethyl-(2,2,3,3-tetramethyl) cyclopropane carboxylate
PCCH	γ-1,2,3,4,5-pentachlorocyclohexane
phthalthrin	3,4,5,6-tetrahydrophthalimidomethyl chrysanthemate
piperonyl butoxide	3,4-methylenedioxy-6-propylbenzyl *n*-butyl diethyleneglycol ether
propanil	3,4-dichloropropionanilide
Prothrin	5-propargyl furfuryl chrysanthemate
pyrethric acid	chrysanthemum dicarboxylic acid monomethyl ester
pyrethrolone	2-pentadienyl-4-hydroxy-3-methyl-2-cyclopenten-1-ene
Ronnel	*O,O*-dimethyl *O*-(2,4,5-trichlorophenyl) phosphorothioate
Sevin	*N*-1-naphthyl methylcarbamate
SKF 525A	2-diethyl aminoethyl diphenyl propylacetate
Sumioxon	*O,O*-dimethyl *O*-(3-methyl-4-nitrophenyl) phosphate
Sumithion	*O,O*-dimethyl *O*-(3-methyl-4-nitrophenyl) phosphorothioate
TCB	1,2,4-trichlorobenzene
tribuphon	*O,O*-dimethyl (1-*n*-butyryloxy-2,2,2-trichloroethyl) phosphonate
Trithion	*O,O*-diethyl *p*-chlorophenyl mercaptomethyl phosphorodithioate

malathion and bromophos. Considerably more effect was noticeable with malathion than with bromophos because DDVP can not only compete for the phosphatases attacking both insecticides but may also inhibit enzymic attacks on the carboxyester side chain of malathion.

Among different classes of insecticide, DDT or lindane could well compete with phosphorothioates for the aerobic lipoxidase/P=S→P=O oxidase/dehydrochlorinase system of the germ.

The paucity of information available on such interactions suggest yet another area of insecticide toxicology that is in need of detailed study.

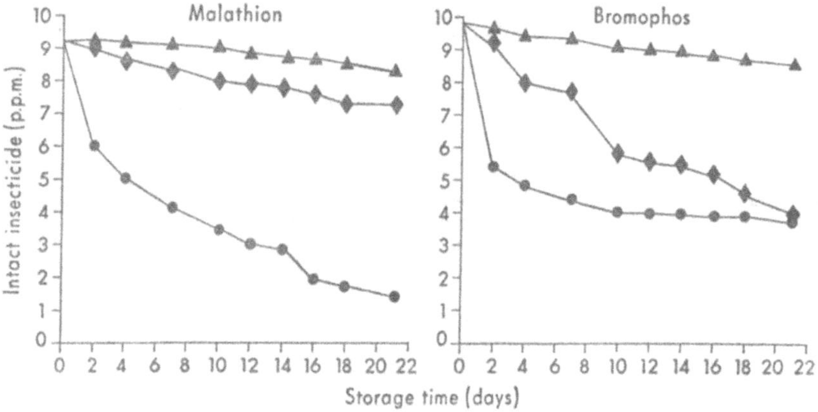

Fig. 10. Breakdown of malathion and bromophos on DDVP-treated wheat: ▲——▲ DDVP-saturated wheat, ◆——◆ wheat treated at ten p.p.m. DDVP, and ●——● untreated wheat

Acknowledgments

The writer would like to thank Messrs. C. E. DYTE, A. A. GREEN, and D. F. HORLER for reading drafts of this manuscript and for their expert criticism thereon.

Summary

The ever-increasing use of grain protectants will undoubtedly result in the incidence and spread of insecticide resistance in stored grain pests which are moved from country to country and commodity to commodity through normal trade channels. The permitted insecticides of practice are still pyrethrins-piperonyl butoxide, lindane, and malathion, although increasing use of Sumithion, DDVP, and bromophos can be expected. Iodophos (or possibly refined Cidial) may also be of potential. Cross-tolerance of resistant insects to many candidate pesticides not yet used in practice may represent a threat to the protective storage of cereal grains and research on possible synergistic combinations and alternative methods of insect control is needed.

Experimental work during treatments of wheat has shown that, although initially all the grains may receive individual doses of insecticide, the distribution through the stored bulk, and amount per grain in any sample taken therefrom, may vary enormously. A few insecticidal grains in a bulk (say five percent) may be sufficient to control insects yet be almost undetectable by chemical assay in a large sample. Spray solutions impinging on grains spread rapidly over the surface and enter the pericarp, the attachment region being the region of most rapid uptake. The routes of entry and subsequent distribution within the wheat kernel are discussed in detail and it is concluded that there is an effective transport system moving the solubilised pesticide, or its metabolic products, from tissue to tissue, even at overall moisture levels as low as seven percent. Factors affecting penetration are also discussed. Similar distribution within a grain of a given pesticide seems to occur a few days after treatment, regardless of temperature, moisture content, age, viability, site of impingement, etc. although there are considerable initial differences, particularly between dust and spray applications.

The main enzymic activity of the wheat grain is found in the globulin fraction of the protein and, at tissue and cellular levels, in the aleurone grains and cytoplasm of the aleurone layers of the endosperm and parenchyma cells of the germ. The oxidases of the seedcoat have little effect on penetrating insecticides and are generally inactive shortly after harvest, but the aerobic lipoxidase-peroxidase-catalase complexes of the germ (in particular) which seem to be more active in older grain ($>$ eight months) have been shown to effect oxidation of organophosphorus and, *in vitro*, of pyrethroid-I type insecticides. These complexes are also implicated in aerobic dehydrochlorination of organochlorine compounds and in reductive dechlorination anaerobically. Decarboxylation and hydrolysis of insecticides, the latter including dealkylation of certain organophosphorus compounds, have been shown to take place in the aleurone and germ tissues. To what extent the parasitic microflora of the cereal grains may be implicated in pesticide metabolism remains to be elucidated positively.

Possible effects on pesticide and endogenous metabolism of various localised natural inhibitors and antioxidants are discussed briefly but are difficult to assess except *in situ;* however, the effects of certain foreign additives such as antimicrobials, fungistats and synergists are a little easier to predict and are discussed in more detail. The relevance of persisting residues of fumigants and other pesticides at time of treatment with contact protectants is also discussed and interactions observed suggest the need for further study.

Résumé*

Le métabolisme des insecticides de contact dans les grains en silos. II. 1966–1969

L'utilisation toujours croissante de produits pour la protection de semences provoquera certainement le développement de la résistance aux insecticides ravageurs de grains emmagasinés qui circulent de

* Traduit par R. Mestres.

pays en pays et d'une denrée à l'autre en suivant les voies commerciales normales.

Les insecticides autorisés sont encore les pyréthrines, le butoxyde de pipéronyle, le gamma HCH et le malathion, bien que l'utilisation croissante du fénitrothion, du DDVP et du bromophos éthyl puisse être attendue. L'iodophos, (ou peut-être le phenthoate) peuvent également être actifs. La tolérance croisée des insectes résistants à de nombreux candidats pesticides non encore utilisés dans la pratique peut être un handicap pour la protection des graines de céréales ensilées. Il est donc nécessaire de rechercher des agents de potentialisation ainsi que d'autres méthodes de lutte contre les insectes.

Les recherches sur les traitements du blé ont montré que la distribution dans la masse stockée et la quantité par graine dans tout échantillon de cette masse peut varier énormément bien qu'au départ tous les grains puissent recevoir des doses individuelles d'insecticides. Quelques graines imprégnées d'insecticide dans la masse, (soit par exemple cinq pour cent) peuvent suffire à la protéger des insectes bien que la teneur moyenne soit indosable dans un échantillon important.

Les solutions pulvérisées frappant les graines s'étendent rapidement sur la surface et pénétrent dans le péricarpe, le point d'attache de la graine étant la région où la prise est la plus rapide. Les voies de pénétration et la distribution qui en résulte dans le grain de blé sont discutées en détail pour conclure qu'il existe un système de transfert efficace faisant circuler le pesticide solubilisé, ou ses métabolites, de tissus en tissus, même à des teneurs en humidité aussi faibles que sept pour cent.

Les facteurs qui affectent la pénétration sont également discutés. Une distribution homogène d'un pesticide donné dans une graine parait survenir quelques jours après le traitement, indépendamment de la température, de l'humidité, de l'âge, de l'état de la graine, du point de premier contact, etc . . . bien qu'il y ait des différences initiales considérables, en particulier entre les applications par poudrage et par pulvérisation.

L'activité enzymatique principale du grain de blé se trouve dans la fraction globuline de la protéine et, au niveau du tissu et de la cellule, dans les grains d'aleurone et le cytoplasme de l'assise protéique et des cellules parenchymateuses du germe.

Les oxydases de l'arille ont peu d'effet sur les insecticides qui pénètrent et sont généralement inactives peu après la récolte, mais on a montré que les complexes aérobies lipoxydase-peroxydase-catalase du germe (en particulier), qui semblent être plus actives dans la graine plus vieille (plus de huit mois), affectent l'oxydation des organophosphorés, et, in vitro des insecticides du type pyréthrine I. Ces complexes interviennent aussi dans la deshydrohalogénation des composés organochlorés et dans la deshalogénation réductive anaérobie. Il a été montré que la décarboxylation et l'hydrolyse des insecticides, cette dernière incluant la desalkylation de certains composés organophosphorés, a lieu dans les grains d'aleurone et les tissus du germe. Il demeure à élucider nettement jusqu'à quel point la microflore parasite des graines de céréales peut être impliquée dans le métabolisme du pesticide.

Les effets possibles sur le métabolisme endogène et celui du pesticide de divers inhibiteurs et antioxydants naturels sont discutés brièvement mais sont difficiles à estimer, sauf in situ; cependant les effets des certains additifs étrangers tels que les anti-microbiens, fongicides et agents de potentialisation sont un peu plus faciles à prévoir et sont discutés avec plus de détails.

La pertinence de résidus persistants de fumigation et d'autres pesticides au moment du traitement avec les produits de protection par contact est également discutée et les intéractions observées suggèrent la nécessité de nouvelles études.

Zusammenfassung*

Der Metabolismus der Kontakt-Insektizide in gelagertem Getreide. II. 1966–1969

Die ständig zunehmende Verwendung von Getreideschutzmitteln wird zweifellos zum Auftreten und zur Verbreitung von Insektizid-Resistenz bei Vorratsschädlingen im Getreide führen, das auf den üblichen Handelswegen von Land zu Land und von Lager zu Lager verschickt wird. Die für die Praxis zugelassenen Insektizide sind noch immer Pyrethrin-Piperonyl-butoxid, Lindan und Malathion. Vermehrte Verwendung von Sumithion, DDVP und Bromophos, vielleicht auch von Jodophos (oder möglicherweise gereinigtem Cidial) ist zu erwarten. Die Kreuz-Toleranz resistenter Insekten für viele in der Praxis noch nicht verwendete Versuchs-Insektizide kann eine Bedrohung für die Getreide-Lagerung darstellen. Nötig ist die Suche nach Synergisten für die Insektizide und nach alternativen Methoden für die Insektenbekämpfung.

Obgleich jedes Weizenkorn,wie experimentell gezeigt wurde, bei der Behandlung zunächst eine individuelle Insektizid-Dosis aufzunehmen vermag, kann die Verteilung des Insektizids in der gesamten gelagerten Masse und die Insektizid-Menge im einzelnen Korn jeder gezogenen Probe stark variieren. Einige Körner (z.B. 5%) in der Gesamtmasse mit genügend hohem Insektizid-Gehalt mögen für die Insektenbekämpfung ausreichen, können aber in einer grossen Getreideprobe analytisch kaum bestimmt werden. Spraylösungen, die auf Getreidekörner auftreffen, verbreiten sich schnell über die Oberfläche und dringen in das Pericarp, die Region der schnellsten Aufnahme, ein. Die Wege des Eintritts und die anschliessende Verteilung

* Übersetzt von H. MAIER-BODE.

innerhalb des Weizenkorns werden im einzelnen diskutiert. Dabei ergibt sich das Vorhandensein eines wirksamen Transportsystems, das den gelösten Wirkstoff oder seine Metaboliten von Gewebe zu Gewebe befördert, selbst bei einem Feuchtigkeitsgehalt von nur 7%. Auch die Faktoren, die das Eindringen beeinflussen, werden diskutiert. Eine gegebene Schädlingsbekämpfungs-Menge scheint einige Tage nach der Applikation innerhalb des Korns unabhängig von Temperatur, Feuchtigkeitsgehalt, Alter, Lebensfähigkeit, der Stelle des Auftreffens u.s.w., in gleicher Weise verteilt zu werden, obgleich kurz nach der Behandlung beträchtliche Unterschiede bestehen, besonders bei Anwendung als Staub oder als Sprühmittel.

Die enzymatische Aktivität findet sich im Weizenkorn hauptsächlich im Globulin-Anteil des Proteins, in den Geweben und Zellen besonders in den Aleuron-Körnern und im Cytoplasma der Aleuronschicht des Endosperms und in den Parenchymzellen des Keimlings. Die Oxydasen der Samenhülle wirken nur wenig auf die eindringenden Insektizide. Sie sind im allgemeinen kurz nach der Ernte unwirksam. Dagegen wirken die aerobischen Lipoxydase-Peroxydase-Katalase-Komplexe des Keimlings, die im älteren Korn (>acht Monate) aktiver zu sein scheinen, oxydativ auf Organophosphorverbindungen und in vitro auf Insektizide von Pyrethroid-1-Typ. Diese Komplexe sind auch an der aerobischen Chlorwasserstoff-Abspaltung und an der anaerobischen reduktiven Chlor-Abspaltung aus organischen Chlorverbindungen beteiligt. Es hat sich gezeigt, dass die Decarboxylierung und die Hydrolyse der Insektizide (letztere einschliesslich Dealkylierung bestimmter Organphosphorverbindungen) im Aleuron und in den Keimlingsgeweben stattfinden. Wie weit die parasitäre Mikroflora der Getreidekörner am Metabolismus der Schädlingsbekämpfungsmittel beteiligt ist, muss noch aufgeklärt werden. Mögliche Wirkungen von lokalisierten natürlichen Inhibitoren und Antioxydantien auf den Endogenen und den Pestizid-Metabolismus werden kurz diskutiert, sind aber in situ schwer abzuschätzen. Indessen sind die Wirkungen bestimmter fremder, z. B. antimikrobieller, fungistatischer und synergistischer Zusatzstoffe etwas leichter zu erkennen. Sie werden im Detail diskutiert. Auch die Bedeutung von Rückständen von Begasungs- und anderen Schädlingsbekämpfungsmitteln die zur Zeit der Behandlung mit Kontaktinsektiziden anwesend sind wird besprochen. Beobachtete Wechselwirkungen weisen auf die Notwendigkeit weiterer Untersuchungen hin.

References

ABDALLAH, M. D.: Interaction of some organophosphorus compounds in susceptible and resistant houseflies. Meded. Landb. Hoogesch. Wageningen **63**, 1 (1963).

AHARONI, A. H., and R. D. O'BRIEN: The inhibition of acetylcholinesterase by anionic organophosphorus compounds. Biochem. **7**, 1538 (1968).

ALESSANDRINI, M. E., V. LEONI, L. DISIMONE, G. IMBROGLINI, and L. ANGELELLI: Andamento dii residui di malathion nei grain trattati in scala semiindustriale e nei relativi prodotti di molitura. Rassegna chimica No. 5, 201 (1968).

ALTSCHUL, A. M., L. Y. YATSU, R. L. ORY, and E. M. ENGLEMAN: Seed proteins. Ann. Rev. Plant Physiol. 17, 113 (1966).

APPLEBAUM, S. W., and A. M. KONIJN: The presence of a *Tribolium* protease inhibitor in wheat. J. Insect Physiol. 12, 665 (1966).

ASHMAN, F., and R. J. KEMPTON: Stability of malathion dilute dust formulations manufactured in Kenya. E. Africa Agr. For. J. 33, 212 (1967).

AYERST, G.: The effects of moisture and temperature on growth and spore germination in some fungi. J. Stored Prod. Research 5, 127 (1969).

BAKRY, N. M. S.: Mode of action of organic thiocyanates as insecticide and carbamate synergists. Dissert. Abstr. 28, 393 (1968).

BARNES, J. R., and J. FELLIG: Synergism of carbamate insecticides by phenyl propynyl ethers. J. Econ. Entomol. 62, 86 (1969).

BATES, A. N., P. S. HEWLETT, and C. J. LLOYD: Synergistic effects of some compounds related to 2-diethylaminoethyl 2,2-diphenyl-n-pentanoate (SKF 525-A) on the insecticidal activity of pyrethrins. J. Sci. Food Agr. 16, 289 (1965).

BERCK, B.: Sorption of phosphine by cereal products. J. Agr. Food Chem. 16, 419 (1968).

BEYNON, K. I., and A. N. WRIGHT: Breakdown of the insecticide "Gardona" on plants and in soils. J. Sci. Food Agr. 20, 250 (1969).

BLINN, R. C., R. W. DORNER, and F. A. GUNTHER: Comparative residual behaviour of pyrethrins and piperonyl butoxide on wheat. J. Econ. Entomol. 52, 703 (1959).

BOWERS, W. S.: Juvenile hormone: Activity of natural and synthetic synergists. Science 161, 895 (1968).

BOWLING, C. C., and W. J. FLINCHUM: Interaction of propanil with insecticide applied as seed treatment on rice. J. Econ. Entomol. 61, 67 (1968).

BROOKE, J. P.: The effect of five methylene dioxyphenyl synergists upon the stability of the pyrethrins. Pyrethrum Post 9, 18 (1967).

BRUINSMA, J.: Plant growth regulators, tools and toys. Meded. Rijksf. Landb. Ghent 31, 343 (1966).

BULL, D. L., and R. L. RIDGWAY: Metabolism of trichlorfon in animals and plants. J. Agr. Food Chem. 17, 837 (1969).

CASIDA, J. E.: Specificity of substituted phenyl phosphorus compounds for esterase inhibition in mice. Biochem. Pharmacol. 5, 332 (1961).

———, J. L. ENGEL, E. G. ESAAC, F. X. KAMIENSKI, and S. KUWATSUKA: Metabolism of methylene-C^{14}-dioxyphenyl compounds in relation to their synergistic action. Science 153, 1130 (1966).

CASTRO, C. E.: The rapid oxidation of iron (11) porphyrins by alkyl halides. A possible mode of intoxication of organisms by alkyl halides. J. Amer. Chem. Soc. 86, 2310 (1964).

CAWLEY, R. W., and T. A. MITCHELL: Inhibition of wheat *a*-amylase by bran phytic acid. J. Sci. Food Agr. 19, 106 (1968).

CHAMP, B. R., and M. J. CAMPBELL-BROWN: Insecticide resistance in Australian *Tribolium castaneum* (Herbst) (Coleoptera, Tenebrionidae). 11. Malathion resistance in Eastern Australia. J. Stored Prod. Research, In press (1970).

———, R. W. STEELE, B. G. GENN, and K. D. ELMS: A comparison of malathion, diazinon, fenitrothion and dichlorvos for control of *Sitophilus oryzae* (L) and *Rhyzopertha dominca* (F) in wheat. J. Stored Prod. Research 5, 21 (1969).

———, and J. CRIBB: Lindane resistance of *Sitophilus sasakii* (Tak) in Queensland. FAO Grain Stor. Newsletter 5, 2 (1963).

CHEN, P. R., W. P. TUCKER, and W. C. DAUTERMAN: Structure of biologically produced malathion monoacid. J. Agr. Food Chem. 17, 86 (1969).

CLARKE, J. H.: Fungi in stored products. Trop. Stored Prod. Inf. **No 15**, 3 (1968).
COGBURN, R. R.: Laboratory tests of five new insecticides as protectants for rough rice. J. Econ. Entomol. **60**, 1286 (1967).
COON, J. M., and J. L. POWERS: Toxicants occurring naturally in foods. Nat. Acad. Sci. Nat. Research Council, Washington, D. C. Publ. 1354, 301 pp. (1967).
DANIELS, D. G. H., and H. F. MARTIN: Antioxidants in oats: Glyceryl esters of caffeic and ferulic acids. J. Sci. Food Agr. **19**, 710 (1968).
DAUSSANT, J., and D. C. ABBOTT: Immunochemical study of changes in the soluble proteins of wheat during germination. J. Sci. Food Agr. **20**, 633 (1969).
DENNIS, D. T., and T. P. COULTATE: Phosphofructokinase, a regulatory enzyme in plants. Biochem. Biophys. Research Communications **25**, 187 (1966).
DRAPER, H. H.: Antioxidants in biological systems. Cereal Sci. Today, abstr. Suppl. (Mar. 1968), joint AACC-ACOS meeting, paper 244 (1968).
DYTE, C. E.: Possible new approach to the chemical control of plant feeding insects. Nature **216**, 298 (1967).
——— Evolutionary aspects of insecticide selectivity. Proc. 5th British Insecticide Fungicide Conf., p. 393 (1969).
———, and D. G. BLACKMAN: Malathion resistance—Rust-red flour beetles. Pest Infestation Research **1968**, p. 48 (1969).
——— ——— The spread of insecticide resistance in *Tribolium castaneum* (Herbst) (Coleoptera, Tenebrionidae). J. Stored Prod. Research, In press (1970).
———, V. J. ELLIS, and C. J. LLOYD: Studies on the contrasting susceptibility of the larvae of two hide beetles (*Dermestes* spp; Coleoptera, Dermestidae) to malathion. J. Stored Prod. Research **1**, 223 (1966).
———, and D. G. ROWLANDS: Some aspects of the specificity of insecticide synergists. Proc. 4th British Insecticide Fungicide Conf., p. 344 (1967).
——— ——— The metabolism and synergism of malathion in resistant and susceptible strains of *Tribolium castaneum* (Herbst) (Coleoptera, Tenebrionidae). J. Stored Prod. Research **4**, 157 (1968).
——— ——— The effects of some insecticide synergists on the potency and metabolism of bromophos and fenitrothion in *Tribolium castaneum* (Herbst) (Coleoptera, Tenebrionidae). J. Stored Prod. Research, **6**, 1 (1970).
——— ———, F. H. FORTESCUE, and M. E. KING: Selectivity and synergism of insecticides: analogues of SKF 525-A and the toxicity of malathion. Pest Infestation Research **1967**, p. 37 (1968).
——— ———, M. E. KING, and J. A. DALY: Selectivity and synergism of insecticides: analogues of SKF 525-A and the toxicity of malathion; metabolism and synergism of bromophos and fenitrothion. Pest Infestation Research **1968**, p. 45 (1969).
——— ———, J. A. DALY, and G. A. LAWRENCE: Synergism of fenitrothion by aliphatic acids. Pest Infestation Research **1969**, p. 44 (1970).
ELLIOTT, M., A. W. FARNHAM, N. F. JANES, P. H. NEEDHAM, and B. C. PEARSON: 5-Benzyl-3-furylmethyl chrysanthemate: A new, potent, insecticide. Nature **213**, 493 (1967).
ENGST, R., and M. KUJAWA: Enzymic decomposition of DDT by a fungus. Nahrung **11**, 751 (1967).
ESAAC, E. G., and J. E. CASIDA: Metabolism in relation to mode of action of methylenedioxyphenyl synergists in houseflies. J. Agr. Food Chem. **17**, 539 (1969).
EVANS, W. C.: Mechanisms of microbial hydroxylations. Biochem. J. **103**, 1P (1967).
FISHBEIN, L., J. FAWKES, H. C. FALK, and S. JORDAN: Thin-layer chromatography

of rat bile and urine following intravenous administration of the pesticidal synergist octachlorodipropyl ether. J. Chromatog. **37**, 256 (1968).

FISHER, N., D. G. REDMAN, and G. A. H. ELTON: Fractionation and characterization of purothionin. Cereal Chem. **45**, 48 (1968).

FONKEN, G. S., M. E. HERR, H. C. MURRAY, and L. M. REINEKE: Microbial hydroxylation of monocyclic alcohols. J. Amer. Chem. Soc. **89**, 672 (1967).

FUKAMI, J. I., and T. SHISHIDO: Nature of a soluble, glutathione-dependent enzyme system active in cleavage of methyl parathion to desmethyl parathion. J. Econ. Entomol. **59**, 1338 (1966).

GARDNER, A. M., J. N. DAMICO, E. A. HANSEN, E. LUSTIG, and R. W. STORHERR: Previously unreported homolog of malathion found in residues on crops. J. Agr. Food Chem. **17**, 1181 (1969).

GARDNER, H. W., and G. E. INGLETT: Correlation of enzymic activities with processing conditions of corn products for CSM. Cereal Sci. Today, Abstr. Suppl. (Mar. 1968), joint AACC-ACOS meeting, paper 181 (1968).

GLYNNE-JONES, G. D., and N. K. SYLVESTER: Pyrethrum as a repellent. 1. Literature review. Pyrethrum Post **8**, 38 (1966).

GRADIDGE, J. M. G., S. L. AGGARWAL, and B. ROBBINS: Treatment of a provender store with phenthoate. Pest Infestation Research **1969**, p. 29 (1970 a).

———, D. R. WILKIN, and V. K. BHARADWAT: Treatment of granaries with iodfenphos. Pest Infestation Research **1969**, p. 28 (1970 b).

GREEN, A. A.: The use of insecticides. Chemy Ind., p. 1452 (1969).

———, and P. S. TYLER: A field comparison of malathion, dichlorvos and fenitrothion for the control of *Oryzaephilus surinamensis* (L) (Coleoptera, Sitranidae) infesting stored barley. J. Stored Prod. Research **1**, 273 (1966).

——— ———, J. KANE, and D. G. ROWLANDS: An assessment of bromophos for the protection of wheat and barley. J. Stored Prod. Research, In press (1970).

———, and D. R. WILKIN: The control of insects in bagged grain by the injection of dichlorvos. J. Stored Prod. Research **5**, 11 (1969).

GREEN, D. E., and I. SILMAN: Structure of the mitochondrial electron transfer chain. Ann. Rev. Plant Physiol. **18**, 147 (1967).

GREEN, J.: The distribution of tocopherols during the life cycles of some plants. J. Sci. Food Agr. **9**, 801 (1958).

GUSS, P. L., T. RICHARDSON, and M. A. STAHMANN: The oxidation-reduction enzymes of wheat. III. Isoenzymes of lipoxidase in wheat fractions and soybeans. Cereal Chem. **44**, 607 (1967).

HADDOW, B. C., and T. G. MARKS: Iodofenphos—A broad review of a promising new insecticide. Proc. 5th British Insecticide Fungicide Conf., p. 531 (1969).

HALL, G. S., and D. L. LAIDMAN: The detection of tocopherols and isoprenoid quinones in the grain and seedlings of wheat; the pattern and control of isoprenoid quinones and tocopherol metabolism in the germinating grains of wheat. Biochem. J. **108**, 465 and 475 (1968).

HAWTHORN, J., and J. P. TODD: Catalase in relation to the unsaturated fat oxidase activity of wheat flour. Chemy Ind., p. 446 (1955).

HERTING, D. C., and E. J. E. DRURY: Alpha-tocopherol content of cereal grains and processed cereals. J. Agr. Food Chem. **17**, 785 (1969).

HEUSER, S. G., and K. A. SCUDAMORE: Formation of ethylene bromohydrin in flour and wheat during treatment with ethylene oxide. Chemy Ind., p. 1054 (1969).

HILL, E. G., and B. S. J. BORDER: Breakdown of malathion during malting processes. Chemy Ind., p. 363 (1967).

———, and R. H. THOMPSON: Pesticide residues in foodstuffs in Great Britain. V. Malathion in imported cereals. J. Sci. Food Agr. **19**, 119 (1968).

HOLLINGWORTH, R. M.: Dealkylation of organophosphorus esters by mouse liver enzymes *in vitro* and *in vivo*. J. Agr. Food Chem. **17**, 987 (1969).

Honold, G. R., G. L. Farkas, and M. A. Stahmann: Oxidation-reduction enzymes of wheat. I. Qualitative investigation of the dehydrogenases. Cereal Chem. **43**, 517 (1966).

——— ——— ——— The oxidation-reduction enzymes of wheat. II. A quantitative investigation of the dehydrogenases. Cereal Chem. **44**, 373 (1967).

———, and M. A. Stahmann: The oxidation-reduction enzymes of wheat. IV. Qualitative and quantitative investigation of the oxidases. Cereal Chem. **45**, 99 (1968).

Hopkins, T. L., and F. W. Knapp: Anticholinesterases in blood and cattle grubs from cattle treated with ronnel. J. Econ. Entomol. **50**, 684 (1963).

Horler, D. F., J. H. Clarke, and E. V. Niles: Field and storage fungi in relation to the breakdown of organophosphorus insecticides on stored grain. Pest Infestation Research **1967**, p. 32 (1968).

———, P. R. Dean, J. H. Clarke, and E. V. Niles: Field and storage fungi in relationship to the breakdown of organophosphorus insecticides on stored grain. Pest Infestation Research **1968**, p. 41 (1969).

Huitson, J. J.: Cereals preservation with propionic acid. Process Biochem. **3**, (1968).

Hurtig, H.: Pesticide residues on cereal grains and their milled fractions. Grain and Cereal Products Sanitation Conf., Manhattan 1966 Proc., 17 pp. (1966).

Innes, J. R. M., B. M. Ulland, M. G. Valerio, L. Petrucelli, L. Fishbein, E. R. Hart, A. J. Pallotta, R. A. Bates, H. L. Falk, J. J. Gart, M. Klein, I. Mitchell, and J. Peters: Bioassay of pesticides and industrial chemicals for tumorigenicity in mice: A preliminary note. J. Nat. Cancer Inst. **42**, 1101 (1969).

Ishida, M.: Comparative studies on BHC metabolizing enzymes, DDT-dehydrochlorinase and glutathione S-transferases. Agr. Biol. Chem. **32**, 947 (1968).

Jacobson, M.: Insect sex attractants. New York-London: Interscience, 154 pp. (1965).

Jotwani, M. G., and P. Sircar: Neem seed as a protectant against stored grain pests infesting wheat seed. Ind. J. Entomol. **27**, 160 (1965).

Kadoum, A. M., and D. W. Lahue: Effect of hybrid, moisture content, foreign material and storage temperature on the degradation of malathion residues in grain sorghum. J. Econ. Entomol. **62**, 1161 (1969).

Kagan, Yu. S., S. I. Fudel-Ossipova, B. J. Khaikina, V. A. Kuzminskaya, and S. D. Koutun: On the problem of the harmful effects of DDT and its mechanism of action. Residue Reviews **27**, 43 (1969).

Kane, J., and S. L. Aggarwal: Protective treatment of bagged barley. Pest Infestation Research **1969**, p. 33 (1970).

———, and A. A. Green: The protection of bagged grain from insect infestation using fenitrothion. J. Stored Prod. Research **4**, 59 (1968).

Karr, D., J. Tweto, and P. Albersheim: S-adenosyl methionine: Methionine methyl transferase from wheat germ. Arch. Biochem. Biophys. **121**, 732 (1967).

Katsuda, Y., T. Chikamoto, H. Ogami, H. Hirobe, and T. Kunishige: New insecticidal chrysanthemic esters. Agr. Biol. Chem. **33**, 1361 (1969).

Katz, M.: Biosynthesis and metabolic degradation of oat coleoptile cell wall constituents. Dissert. Abstr. **B.27**, 3821 (1967).

Kenefick, D. G., and J. B. Hanson: The site of oligomycin action in corn mitochondria. Biochem. Biophys. Research Communications **24**, 899 (1966).

Kent, N. L.: Subaleurone endosperm cells of high protein content. Cereal Chem. **43**, 585 (1966).

Kies, M. W., J. L. Haining, E. Pistorius, D. H. Schroeder, and B. Axelrod: On the question of the identity of soybean lipoxidase and carotene oxidase. Biochem. Biophys. Research Communications **36**, 312 (1969).

Klun, J. A., and J. F. Robinson: Concentration of two 1,4-benzoxazinones in

Dent corn at various stages of development of the plant and its relation to resistance of the host plant to the European corn borer. J. Econ. Entomol. 62, 214 (1969).

KOHN, G. K.: Attenuation of pesticide residues on seeds. U. S. Pat. 3,305,441 (Cl.167-38) Feb. 21. (1967).

——— Attenuation of pesticide residues on seeds. Residue Reviews 29, 3 (1970).

KRETOVITCH, V. L.: Physiologico-biochemical bases for the storage of wheat. Akad. Nauk. SSSR. Inst. Biokhim., Moscow-Leningrad 1 (1945).

KRUGER, J. H., and R. TKACHUK: Wheat alpha-amylase: Isolation and properties. Cereal Sci. Today, Abstr. Suppl. (Mar. 1968) joint AACC-ACOS meeting, paper 179 (1968).

KUGE, T., and K. TAKEO: Complex formation of starch with organic substances. Affinity observed by gas chromatography. Agr. Biol. Chem. 31, 257 (1967).

KURASAWA, H., T. HAYAKAWA, and S. MOTODA: Biosynthesis of inositol by an enzyme system in the rice seed. Agr. Biol. Chem. 31, 382 (1967).

LAHUE, D. W., and C. C. FIFIELD: Evaluation of four inert dusts on wheat as protectants against insects in a small bin. U. S. Dept. Agr., Agr. Research Service, Marketing Research Rept. 780, p. 1 (1967).

LAWLER, P. D., and L. J. ROGERS: Effect of DDT on photosynthesis in certain varieties of barley. Nature 215, 1515 (1967 a).

——— ——— DDT as an inhibitor of photophosphorylation in certain varieties of barley. Biochem. J. 105, 11P (1967 b).

LAWS, E. Q.: Residues from agricultural pesticide treatment. Proc. 1st Internat. Congress Food Sci. Technol. 3 London: Gordon & Broach (1965).

LEARMOUTH, E. M., and J. C. WOOD: A trypsin inhibitor in wheat flour. Chemy Ind., p. 1969 (1960).

LEMON, R. W.: Laboratory evaluation of malathion, bromophos and fenitrothion for use against beetles infesting stored products. J. Stored Prod. Research 2, 197 (1967 a).

——— Laboratory evaluation of some additional organophosphorus insecticides against stored product beetles. J. Stored Prod. Research 3, 283 (1967 b).

LIEBERMAN, M., and J. E. BAKER: Respiratory electron transport. Ann. Rev. Plant Physiol. 16, 343 (1965).

LINDGREN, D. L., W. B. SINCLAIR, and L. E. VINCENT: Residues in raw and processed foods resulting from post-harvest insecticide treatments. Residue Reviews 21, 1 (1968).

LINKO, P.: Water content and metabolism of wheat during short storage and germination. Suomal. Tiedeakat. Toim. Ser. A. 2. Chimica 98, 69 pp. (1960).

LITTLE, C., and P. J. O'BRIEN: The effectiveness of a lipid peroxide in oxidizing protein and non-protein thiols. Biochem. J. 106, 419 (1966).

LLOYD, C. J., and J. FIELD: Tests of newer insecticides: A synthetic pyrethroid NRDC 104. Pest Infestation Research 1967, p. 41 (1968).

——— ——— Tests of newer insecticides: synthetic pyrethroids. Pest Infestation Research 1968, p. 46 (1969).

———, D. G. ROWLANDS, and J. FIELD: Synthetic pyrethroids. Pest Infestation Research 1969, p. 45 (1970).

LONGONI, A., and G. MICHIELI: Initial and residual activity of cidial (phenthoate) on insects affecting stored products as compared with malathion and lindane. Proc. 5th Internat. Congress Plant Prot., Vienna (1967).

LOOMIS, W. F.: Fed. Proc., Fed. Amer. Soc. Expt. Biol. 8, 220 (1949); cited by Slater, E. C.: Uncouplers and inhibitors. In HOCHESTER, R. M., and J. H. QUASTEL (Eds.): Metabolic inhibitors 2, p. 503. New York: Academic Press (1963).

LUDWICZAK, R. S., A. RYKOWSKI, and I. ZYCZYNSKA: Neutral constituents of wheat gluten. Rocz. Chem. 41, 1639 (1967); through Chem. Abstr. 68, 27525r (1968).

L'VOVA, L. S.: The effect of moulds on the enzymic activity, respiration and viability of stored grains. Prinkl. Biokhim. Mikrobiol. **4**, 218 (1968); through Chem. Abstr. **69**, 770a (1968).

MACKO, V., G. K. HONOLD, and M. A. STAHMANN: Soluble proteins and multiple enzyme forms in early growth of wheat. Phytochem. **6**, 465 (1967).

MCGAUGHEY, W. H.: Malathion residues on rice. Rice J. **72**, 3 (1969).

MACLEOD, A. M., and G. H. PALMER: Giberellins from barley embryos. Nature **216**, 1342 (1967).

MACMASTERS, M. M., D. BRADBURY, and J. J. C. HINTON: Microscopic structure and composition of the wheat kernel. In HLYNKA, I.: Wheat, chemistry and technology. Amer. Assoc. Cereal Chemists, Minneapolis, Minn. (1964).

MACRAE, I. C., K. RAGHU, and T. F. CASTRO: Persistence and biodegradation of four common isomers of benzene hexachloride in submerged soil. J. Agr. Food Chem. **15**, 911 (1967).

——— ——— and E. M. BAUTISTA: Anaerobic degradation of the insecticide lindane by *Clostridium* spp. Nature **221**, 859 (1969).

MAIN, A. R.: The differentiation of the A-type esterases in sheep serum. Biochem. J. **75**, 188 (1960).

MAJOR, W., and E. H. ROBERTS: Dormancy in cereal seeds. II. Nature of the gaseous exchange in imbibed barley and rice seeds. J. Expt. Bot. **19**, 90 (1968).

MANDEL'BAUM, Y. A., L. A. DAVYDOV, N. N. MEL'NIKOV, Z. M. BAKANOVA, and I. A. PIL'-MENSHTEIN: Substituted phenoxy and phenylthiodiethyl sulphides as synergists for organophosphorus insecticides. Khim. Pr. **44**, 341 (1968); through Chem. Abstr. **69**, 34969n (1968).

MATHLEIN, R.: Laboratory trials with chemical repellents against stored product pests. Meddn. St. Växtsk. Anst. **13**, 447 (1967).

MATSUMURA, F., and K. SAKAI: Degradation of insecticides by esterases of the American cockroach. J. Econ. Entomol. **61**, 598 (1968).

MENZER, R. E., and J. E. CASIDA: Nature of toxic metabolites found in mammals, insects, and plants from 3-(dimethoxyphosphinyloxy)-*N*,*N*-dimethyl-*cis*-crotonamide and its *N*-methyl analog. J. Agr. Food Chem. **13**, 102 (1965).

MEREDITH, O. G., and J. J. WREN: Determination of molecular weight distribution in wheat flour proteins by extraction and gel filtration in a dissociating medium. Cereal Chem. **43**, 169 (1966).

MEREDITH, W. O. S.: 1. Free and bound beta amylase in cereal species. 11. Distribution of protein and free and bound amylases in cereal species. Amer. Soc. Brewing Chem., Proc., p. 27 (1966).

METCALF, R. L.: Mode of action of insecticide synergists. Ann. Rev. Entomol. **12**, 229 (1967).

MIKKELSEN, D. S.: Germination inhibitors as a possible factor in rice dormancy. Internat. Rice Comm. Newsletter Spec. Issue "Symp. Prob. Devt. Ripening Rice Grain," p. 32 (1967).

MILNE, D. L.: Food preservatives as insecticides against *Tribolium confusum* Du Val, *T. castaneum* Herbst, and *Lasioderma serricorne* F. S. African J. Agr. Sci. **7**, 79 (1964).

MISKUS, R. P., D. P. BLAIR, and J. E. CASIDA: Conversion of DDT to DDD by bovine rumen fluid, lake water and reduced porphyrins. J. Agr. Food Chem. **13**, 481 (1965).

MITSUDA, H., K. YASUMOTO, and K. IWAMI: Volatile compounds in rice bran. Agr. Biol. Chem. **4**, 453 (1968).

MIYAMOTO, J., K. KITAGAWA, and Y. SATO: Metabolism of organophosphorus insecticides by *Bacillus subtlis* with special emphasis on Sumithion. Jap. J. Expt. Med. **36**, 211 (1966).

MURATA, T., N. NAKAYAMA, Y. TANAKA, and T. AKAZAWA: Enzymic mechanisms of starch synthesis in ripening rice grain. V. Iodoacetate inhibition of granular starch synthetase activity. Arch. Biochem. Biophys. **123**, 97 (1968).

NEALSON, K. H., and L. D. GARBER: An electrophoretic survey of esterases, phosphatases and leucine aminopeptidase in mycelial extracts of species of *Aspergillus*. Mycologia **59**, 330 (1967).

NEWMAN, A. A.: The occurrence, genesis and chemistry of the phenolic methylenedioxy ring in nature (2 parts). Chem. Prod. **25**, 115 and 161 (1962).

NIP, W. K., and E. E. BURNS: Pigment characterization in grain sorghum. Cereal Sci. Today, Abstr. Suppl. (Mar. 1968) joint AACC-ACOS meeting, Paper 87 (1968).

NOVOTEL'NOV, N. V., K. K. GORBATOVA, G. V. LUSHNIKOVA, and N. A. BELOVA: Organic acids of wheat bran and their influence on fungal amylase. Izv. Vysshikh. Ucheln. Zavedenii, Piskehevaga Tekhnol., p. 33 (1966).

PARK, J. H., and D. E. KOSHLAND: The hydrolytic activity of glyceraldehyde-3-phosphate dehydrogenase. J. Biol. Chem. **233**, 986 (1958).

PARKIN, E. A.: The onset of insecticide resistance among field populations of stored product insects. J. Stored Prod. Research **1**, 3 (1965).

———, and R. FORSTER: Increased resistance of stored product insects to insecticides: the rust-red flour beetle and malathion. Pest Infestation Research **1961**, p. 84 (1962).

PEERS, F. G.: Germination inhibitory substances in oat husk. W. African J. Biol. Chem. **2**, 9 (1958).

PIXTON, S. W., and S. T. HILL: Long term storage of wheat. II. J. Sci. Food Agr. **18**, 94 (1967).

PLAPP, F. W., and H. H. C. TONG: Synergism of malathion and parathion against resistant insects: phosphorus esters with synergistic properties. J. Econ. Entomol. **59**, 11 (1966).

———, and T. M. VALEGA: Synergism of carbamates and organophosphate insecticides by non-insecticidal carbamates. J. Econ. Entomol. **60**, 1094 (1967).

POMERANZ, Y.: Role of the lipid fraction in the growth of cereals and in their storage and processing. Wallerstein Lab. Communications **29**, 17 (1966).

PRENTICE, N., W. C. BURGER, and J. KASTENSCHMIDT: The distribution of acidic and neutral peptidases in barley and wheat kernels. Physiol. Plant. **20**, 361 (1967).

REDMAN, D. G., and N. FISHER: Purothionin analogues from barley flour. J. Sci. Food Agr. **20**, 427 (1969).

REED, D. J.: Mode of action of hydrazine growth retardants. Plant Physiol. **40**, 59 (1965).

REIFER, I., G. MUSZYNSKA, and E. BER: Natural inhibitors in higher plants. Bull. Acad. Pol. Sci. Ser. Sci. Biol. **16**, 9 (1968).

RIBBONS, D. W.: Microbial degradation of aromatic compounds. Ann. Rept. Chem. Soc. **2**, 446 (1965).

ROBERTS, E. H.: Seed dormancy and oxidation processes. Symp. Soc. Expt. Biol. **23**, 161 (1969).

ROHRLICH, M., and F. MEUSER: Enzymic metabolism of moist, phosphine-treated grain. Ber. Getreidechem.-Tag., Detmold, p. 71 (1969).

———, and TH. NIEDERAVER: Lipid-protein complexes in cereals. 1. Isolation of lipoproteins from wheat, rye and oats. Fette, Seifen, Anstricken **69**, 63 (1967).

ROONEY, L. W.: The lipids of sorghum grain. 1. Characterisation of free and bound lipids from selected grain sorghum varieties. Cereal Sci. Today, Abstr. Suppl. (Mar. 1968) joint AACC-ACOS meeting (1968).

ROSENBERG, H., and J. M. LA NAUZE: Metabolism of phosphonates by microorganisms. The transport of aminoethyl phosphoric acid in *Bacillus cereus*. Biochem. Biophys. Acta. **141**, 79 (1967).

ROTHE, M., T. STOECKEL, and R. HECKEL: Lipase activity of rye. Ernährungsforschung **12**, 265 (1967); through Chem. Abstr. **67**, 88300q (1967).

Rowlands, D. G.: The metabolism of bromophos in stored wheat grains. J. Stored Prod. Research **2**, 1 (1966 a).

——— The activation and detoxification of three organic phosphorothionate insecticides applied to stored wheat grains. J. Stored Prod. Research **2**, 105 (1966 b).

——— The metabolism of contact insecticides in stored grains. Residue Reviews **17**, 105 (1967).

——— The metabolism of DDT in stored wheat grains. J. Stored Prod. Research **4**, 183 (1968 a).

——— Fate of insecticide residues in stored grain: Allethrin, DDT, dichlorvos, fenthion, SD 8447. Pest Infestation Research **1967**, p. 30 (1968 b).

——— Effect of prior fumigation with methyl bromide on the desmethylation of organophosphorus insecticides. Pest Infestation Research **1968**, p. 40 (1969).

——— The metabolic fate of dichlorvos on stored wheat grains. J. Stored Prod. Research, **6**, 19 (1970 a).

——— Metabolic activity towards insecticide substrates in the germ of wheat grains. In press (1970 b).

——— Comparison of rates of degradation of dichlorvos on wheat and barley. Pest Infestation Research **1969**, p. 37 (1970 c).

———, and J. H. Clarke: Metabolism of insecticides in stored cereals: Effect of bromophos residues on grain fungi. Pest Infestation Research **1967**, p. 32 (1968).

———, and D. F. Horler: Penetration of malathion into wheat grains. Proc. 4th British Insecticide Fungicide Conf. **1**, 331 (1967).

———, and C. J. Lloyd: DDT metabolism in susceptible and pyrethrin-resistant *Sitophilus granarius* (L) (Coleoptera, Curculionidae). J. Stored Prod. Research **5**, 4132 (1969).

———, and R. K. Sharma: Metabolism of insecticides in stored cereals: DDT. Pest Infestation Research **1967**, p. 30 (1968).

——— ———, and P. R. Dean: Nature of the oxidative activity in the seed coat. Pest Infestation Research **1967**, 29 (1968).

Rowsell, E. V., and M. Y. Ali Khan: Respiration and dehydrogenase activities of wheat aleurone. Biochem. J. **100**, 56P (1966).

——— ———, and S. M. Haqjoo: Gibberellic acid stimulation of wheat endosperm hydrolysis: Dependence on oxygen. Biochem. J. **100**, 55P (1966).

Sacher, R. M., R. L. Metcalf, and T. R. Fukuto: Propynyl naphthyl ethers as selective carbamate synergists. J. Agr. Food Chem. **16**, 779 (1968).

Schesser, J. H.: A dichlorvos-malathion mixture for insect control in empty rail cars. Amer. Miller and Processor, p. 1, Sept. (1967).

Schroeder, H. W.: Sodium propionate and infrared drying for control of fungi infecting rough rice. Phytopathol. **54**, 858 (1964).

Scott, A. I.: Some natural products derived by phenol oxidation. In Taylor, W. I., and A. R. Battersby (Eds.): Organic substances of natural origin, Vol. 1, Oxidative coupling of phenols. London: Edward Arnold (1967).

Shyamala, G., B. M. Kennedy, and R. L. Lyman: Trypsin inhibitors in whole wheat flour. Nature **192**, 360 (1961).

Sloane, N. H., and M. Heinemann: Hydroxymethylation of the benzene ring. II. Sequential formation of *p*-hydroxyaniline via hydroxymethylation of aniline by micro-organisms. Biochem. Biophys. Acta. **141**, 47 (1967).

Slover, H. T., J. Lehmann, and R. J. Valis: Vitamin E in foods; Determination of tocols and tocotrienols; Tocopherols in wheat and wheat products. Cereal Sci. Today, Abstr. Suppl. (Mar. 1968) joint AACC-ACOS meeting, Papers 243 and 253 (1968).

Sobolev, A. M., and M. A. Rodionova: Phytin synthesis by aleurone grains in maturing seeds. Fiziol. Rast. **13**, 1090 (1966).

SOLOWAY, A. H.: Aromatic hydroxylation—A new biosynthetic mechanism. J. Theoret. Biol. **13**, 100 (1966).
SPEIRS, R. D., L. M. REDLINGER, and H. P. BOLES: Malathion resistance in the red flour beetle. J. Econ. Entomol. **60**, 1373 (1967).
STANIC, V., and A. SHULOV: Olfactory responses of *Trogaderma granarium adults* (Col., Dermest.) to odours of wheat, faeces and some faecal components. J. Stored Prod. Research **5**, 299 (1969).
STAUFFER, C. E., and R. L. GLASS: The glycerol ester hydrolases of wheat germ. Cereal Chem. **43**, 644 (1966).
STENERSEN, J.: Degradation of P^{32}-bromophos by micro-organisms and seedlings. Bull. Envir. Contamination Toxicol. **4**, 104 (1969 a).
——— Demethylation of the insecticide bromophos by a glutathione dependent liver enzyme and by alkaline hydrolysis. J. Econ. Entomol. **62**, 1043 (1969 b).
STREET, J. C.: Methods of removal of pesticide residues. Canad. Med. Assoc. J. **100**, 154 (1969).
STRONG, R. G., and D. E. SBUR: Protective sprays against internal infestations of grain beetles in wheat. J. Econ. Entomol. **57**, 544 (1964).
——— ——— Evaluation of insecticides for control of stored product insects. J. Econ. Entomol. **61**, 1034 (1968).
——— ———, and R. G. ARNDT: Influence of formulation in the effectiveness of malathion, methoxychlor and synergised pyrethrum sprays for stored wheat. J. Econ. Entomol. **54**, 489 (1961).
——— ———, and G. J. PARTIDA: The toxicity and residual effectiveness of malathion and diazinon used for protection of stored wheat. J. Econ. Entomol. **60**, 500 (1967).
SUGIMOTO, A.: Bioassay of α-BHC residues in rice grains and straws. Jap. J. Applied Entomol. Zool. **10**, 156 (1966).
SUGIYAMA, H., and H. SHIGEMATSU: Enzymic dehydrochlorination of trichlorfon by the digestive juice of the silkworm *Bombyx mori* (L). Botyu Kagaku **34**, 79 (1969).
SURTEES, G.: Site of damage to whole wheat grains by five species of stored product insects. Entomol. Mon. Mag. **99**, 178 (1964).
TANAKA, Y., S. MINAGAWA, and T. AKAZAWA: Association of enzyme proteins with starch granules in rice grains. Die Starke **19**, 206 (1967).
TAPPEL, A. L.: The mechanisms of the oxidation of unsaturated fatty acids catalysed by hematin compounds. Arch. Biochem. Biophys. **44**, 378 (1953).
TAYLOR, H. F., and T. A. SMITH: Production of plant growth inhibitors from xanthophylls: A possible source of dormin. Nature **215**, 1513 (1967).
THOMPSON, J. F.: Sulfur metabolism in plants. Ann. Rev. Plant Physiol. **18**, 59 (1967).
THOMPSON, R. H.: A review of the properties and usage of methyl bromide as a fumigant. J. Stored Prod. Research **1**, 353 (1966).
———, and E. G. HILL: Pesticide residues in foodstuffs in Great Britain. XI. Further studies of malathion in imported cereals. J. Sci. Food Agr. **20**, 293 (1969).
TOMIZAWA, C., and T. SATO: Metabolic fate of malathion and methyl parathion in rice plant. Japan J. Applied Entomol. Zool. **6**, 70 (1962).
TYLER, P. S., and A. A. GREEN: The effectiveness of fenitrothion and malathion as grain protectants under severe practical conditions. J. Stored Prod. Research **4**, 119 (1968).
———, and D. G. ROWLANDS: Protective treatment of grain. Pest Infestation Research **1967**, p. 25 (1968).
UCLAF, R.: Bioallethrin. Procida, Puteaux (France), 39 pp. (1968).
UN/FAO: World crop statistics: Area, production and yield. 1948–1964. Rome, 458 pp. (1966).

——— The state of food and agriculture 1969. Rome, 198 pp. (1969).
——— FAO/WHO Expert Committee on Food Additives: Specifications for identity, purification and toxicological evaluation of some antimicrobials and antioxidants. Rept. Meeting Geneva (Dec. 1964). Nutrition Studies Rept. No. 38A, 90 pp. (1965).
——— FAO Working Party on Pesticide Residues/WHO Expert Committee on Pesticide Residues: Report of Geneva meeting (Dec. 1968). Agricultural Studies No. 78, 40 pp.; also Monographs: 1968 Evaluation of some pesticide residues in foods. FAO/PL: 1968/M/9/1; WHO/FOOD ADD./69.38 (1969).
Varner, J. E.: Seed development and germination. In Bonner, J. M., and J. E. Varner (Eds.): Plant biochemistry. New York-London: Academic Press (1967).
Verma, J. P.: Studies on the enzyme make-up of alternaria. Mycopathol. et Mycol. Appl. **23**, 131 (1964).
Vikhanskii, Yu. D.: Comparative toxicity of organophosphorus pesticides for grain and grain product pests. Through Chem. Abstr. **67**, 107639a (1967).
Weaving, A. J. S., and N. K. Sylvester: Pyrethrum as an insect repellent 11. Laboratory technique for its evaluation and influence of formulation on persistence. Pyrethrum Post **9**, 31 (1967).
Weber, E. J.: Lipids of maturing grain corn (Zea mays). Cereal Sci. Today, Abstr. Suppl. (Mar. 1968) joint AACC-ACOS meeting, Paper 127 (1968).
Weinbach, E. C., and J. Garbus: Mechanism of action of reagents that uncouple oxidative phosphorylation. Nature **221**, 1016 (1969).
Whetstone, R. R., D. D. Phillips, Y. P. Sun, L. F. Ward, and T. E. Shellenberger: 2-Chloro-1-(2,4,5-trichlorophenyl) vinyl dimethyl phosphate, a new insecticide with low toxicity to mammals. J. Agr. Food Chem. **14**, 352 (1966).
Wilkin, D. R., A. A. Green, J. M. G. Gradidge, J. Kane, S. L. Aggarwal, K. Thomas, D. F. Horler, and J. S. Clarke: Insecticidal treatment of grain: Admixture of iodfenphos. Pest Infestation Research (1969), p. 31 (1970).
———, and K. Thomas: Treatment of grain with propionic acid. Pest Infestation Research (1969), p. 32 (1970).
Wilkinson, C. F.: The relationship of structure and mode of action of synergists for carbamate insecticides. Ph.D. Thesis, Univ. Calif., Riverside (1965).
——— The role of insecticide synergists in resistance problems. World Rev. Pest Control Research **7**, 155 (1968).
———, and L. J. Hicks: Microsomal metabolism of the 1,3-benzodioxole ring and its possible significance in synergistic action. J. Agr. Food Chem. **17**, 829 (1969).
Williams, C. M.: Juvenile hormone insecticides. Proc. Internat. Symp.: New perspectives in the control of injurious insects, Rome (Sept. 1968).
Wilson, B. J.: Fungal toxins. In Coon and Powers (1967) qv.
Winteringham, F. P. W., and A. Harrison: The sorption of methyl bromide by wheat. J. Soc. Chemy Ind. **65**, 140 (1946).
Whitting, L. A.: Tocopherol destruction and peroxide formation with ethyl oleate, linoleate, linolenate and arachidoleate. Cereal Sci. Today, Abstr. Suppl. (Mar. 1968) joint AACC-ACOS meeting, paper 242 (1968).
Wolff, I. A.: Seed lipids. Science **154**, 1140 (1966).
Yamamoto, I.: Mode of action of natural insecticides. Residue Reviews **25**, 161 (1969).
———, and J. E. Casida: *O*-Demethyl pyrethrin II analogs from oxidation of pyrethrin I, allethrin, dimethrin and phthalthrin by a housefly enzyme system. J. Econ. Entomol. **59**, 1542 (1966).
Ysauno, M., S. Hirakoso, M. Sasa, and M. Uchida: Inactivation of some organo-

phosphorus insecticides by bacteria in polluted water. Jap. J. Expt. Med. 35, 545 (1965).
ZAKI, M. M., A. EL-GHAR, and A. A. HAFEZ: The effect of certain insecticides on the viability of stored wheat grain. Ann. Agr. Sci., Univ. A'in Shams 8, 25 and 37 (1966).
ZUBAIDOV, U.: Seed globulins of some sorghum varieties. Dokl. Akad. Nauk Tadzk. SSR **11**, 70 (1968); through Chem. Abstr. **69**, 8867h (1968).

Manuscript received 2 February 1970; accepted 21 April 1970.

The role of pesticides in pollution management

By

KENNETH C. WALKER*

Contents

I. Introduction

In considering the topic of pollution management and the role of pesticides in pollution management one needs first to establish several points. What is pollution? Can insecticides in fact become pollutants? What is the extent of the concern in this area? What is being done and what more can be done to help alleviate the problem, if in fact a problem does exist?

There have been several rather exhaustive studies made of pollution in general and of pesticides as a potential pollutant. One of the better known studies was that of the Environmental Pollution Panel of the United States' President's Science Advisory Committee. This report was addressed to the quality, or the lack of quality, of our environment. At the outset of the report, the following definition is presented for environmental pollution:

"Environmental pollution is the unfavorable alteration of our surroundings, wholly or largely as a byproduct of man's actions, through direct or indirect effects of changes in energy patterns, radiation levels, chemical and physical constitution and abundance of organisms. These changes may affect man directly, or through his supplies of water and of agricultural and other biological products, his physical objects or possessions, or his opportunities for recreation and appreciation of nature."

---

* Assistant to Deputy Administrator, Agricultural Research Service, *U. S. Department of Agriculture*, Washington, D. C. 20250.

This is a rather impressive and all-encompassing definition of environmental pollution. It leaves little that cannot be included under such a definition. Pesticides, as a pollutant, readily fit into this definition. Pesticides can affect man directly, or through his supplies of water, and of agricultural and other biological products.

On March 8, 1969 President Johnson, in his message to Congress entitled "To Renew A Nation," charged the Secretary of Agriculture with conducting a government-wide review of agricultural waste problems. Accordingly, a study was undertaken in cooperation with the Office of Science and Technology. Seven major pollutants and the general subject of air pollution were studied. In the report of this study, pesticides were defined as a pollutant when ". . . they, their metabolites, or their degradation products remain in the environment after the desired purpose has been accomplished or when they reach some part of the environment other than the intended target." Some pesticides, or their degradation products do remain after they have accomplished their intended purpose. There is evidence of pesticide residues in parts of the environment other than the intended target. By this definition, pesticides can and do become environmental pollutants.

II. Pesticides in food and the environment

In man's immediate environment we know that pesticides do occur in our food supply. The well-known "market basket" studies of the Food and Drug Administration have given us a good measurement of the level of pesticides ingested by our largest eater—the 16 to 19 year-old male. In general, the levels of pesticides, when they are present, are approximately one-tenth of what has been established as the allowable daily intake, or ADI. This is the calculated amount that man can ingest daily for his lifetime with no measurable untoward effects. The fact that we are well below these figures is reassuring and continued efforts are underway to reduce this intake even further. It is of interest that similar studies have been undertaken in other countries. The final evaluation of any pesticide residue control program can only be attained by such studies.

A national water quality network permits the periodic sampling and analysis of our major water resources. This program has been underway for approximately ten years. Low levels of certain pesticides have been found in grab samples as well as samples collected by means of carbon filters. In the order of frequency of occurrence, dieldrin, endrin, DDT, and DDE have been found in all major river basins. Not all of these pesticides have been found in all of the water sources sampled. Generally the pesticide, when found, was at a level of less than a part per billion. Evidence to date does not indicate that the levels of pesticides in surface waters present an acute toxicity hazard to man. However, not enough is known at present about the effects of long-term low level environmental contamination, either on man, aquatic organisms including fish, or on other portions of the environment.

The occurrence of residues in fish and wildlife has been reported and has been a subject of controversy for several years. A number of reports have been issued on studies made on wildlife in areas treated with pesticides and, as would be expected, accumulations similar to those experienced in domestic animals under like conditions were found.

Because of their long persistence and tendency to concentrate in living organisms, residues of several widely used chlorinated hydrocarbon insecticides have been found in practically all animal tissues analyzed. Among these are specimens from remote areas where pesticides have not been known to be used. The residue levels in the animals vary from a fraction of a part per million to as high as several thousand parts per million. There appears to be a general correlation between the pesticide residue level and the position of the animal in the food chain.

Some species of fish and wildlife are highly sensitive to certain pesticides. As a general rule crustaceans, mollusks, and fish are most sensitive, followed by amphibians, reptiles, birds, and mammals. Effects of pesticides other than acute lethality have been reported. Reproduction losses of certain birds and fish have been associated with pesticide residues in the eggs.

The sampling and analysis of air for pesticide residues is complicated by the co-existence of the pesticides in the particulate, vapor, and aerosol phases. The mechanisms for entrapment of the vapor and aerosol phases is still not completely satisfactory. Aerial drift of pesticides is widely recognized. The extent of movement of particulates can be illustrated by the example of a dust storm occurring in the Southwest citrus area carrying pesticide particulates through upper atmospheric stratas. Rainfall carried these particulates downward and caused an increase in the pesticide levels of the rainwater as far east as Cincinnati and the Ohio River Valley.

Our soils are among our most valuable natural resources. Our agricultural productivity and economic wealth result in a large part from the vast areas of fertile land. Although it is estimated that less than five percent of our total land area in the 48 contiguous States receives insecticide applications, many of our most productive soils are treated several times each year. It was not too long ago that the persistent nature of some of our insecticides was considered as a virtue. For some uses persistence is still a virtue and for others it is now a drawback.

The Department of Agriculture conducts a nationwide soil monitoring program as part of the National Monitoring Network coordinated by the Monitoring Subcommittee of the Federal Committee on Pest Control. Extensive sampling of well-defined areas where pesticides have been used for many years indicate that there is no progressive buildup or accumulation of pesticides in the soil. In general, the amount of a particular pesticide found is roughly equal to the amount applied during the current or previous year. While these data point to a lack of buildup, they do not elucidate the mechanism of residue loss from the soils. Much additional work must be done to develop the pathways of loss from the soil.

The American people, and people of other nations, have a growing concern about the impact of technological changes on human and social welfare—on the natural resources upon which we all must depend and which for the most part constitute the physical environment in which we live. Man is concerned regarding his environment.

We have come to know that advances in technology making possible new things for better living have often been accompanied by alterations in our environment, that not all of these changes are desirable, many are unwanted, and some are dangerous. The introduction of pesticides into the environment by man has produced desirable and, in some cases, undesirable effects.

III. Public and governmental reactions

To many people these matters have become burning issues. They are vocal in their inquiries, demands, pleas, and, in rare instances, in their threats. These matters of concern have been brought to the attention of local, State, and National Governments; agencies dealing with natural resources; and farmers, foresters, and industries who utilize the materials in question or who utilize the products of our natural resources. These matters of concern are real both to those who voice their concern and to those to whom the concern is expressed. These concerns for the environment are becoming to be shared by a growing percentage of the total citizenry. That is good, for in the end the solution of these and related problems may well require the combined efforts of our Nation.

How concerned are people? The American Institute of Public Opinion, or better known to some of us as the Gallup Poll, recently conducted a cross-sectional study of the Nation's adults regarding their concern for the environment. Half of all the people interviewed stated that they were "deeply concerned" about the effect of air pollution, water pollution, soil erosion, and destruction of wildlife on our natural surroundings. Only 12 percent said they were "not very concerned." The extent of the depths of their concern can perhaps best be measured by the fact that three out of four who were concerned stated that they would be willing to pay additional taxes for the purpose of combating environmental deterioration. This is a surprise finding in these days of consciousness regarding the taxes one pays.

This same poll indicated that seven percent of those interviewed were concerned with the impact of pesticides on our wildlife. We should take little comfort from the relatively low percentage of positive responses. The fact that there is a concern expressed by a segment of the population regarding the impact of pesticides on a portion of our environment should cause us to pause and consider their concern.

The public's concern on the environment is also reflected in the actions of the people's elected representatives in the State and National legislative bodies. There are few other issues at the present time that command so much attention. Approximately 12 State legislatures have or have had some proposed legislative action regarding pesticides. Most of the proposed action was aimed at curtailing or banning DDT. More than 20 proposals have been introduced in the National Congress to provide for conservation of our natural resources and improvement of our environment. Some of these bills, like those at the State level, are aimed at stronger control of pesticides.

On May 29, 1969 President Nixon, by Executive Order Number 11472 established the Environmental Quality Council. This Council has a status similar to that of the National Security Council. Dr. Lee A. DuBridge, the President's Science Advisor, will serve as Executive Secretary of the Council. Dr. DuBridge is quoted as stating that the problems of DDT residues in the environment will be high on the agenda of the new Council. Secretary of Agriculture Hardin is a member of the new Council along with several other Secretaries. Our Department will participate fully in the deliberations of this Council.

Those who follow the news will be well aware that DDT has been very much in the news. Depending on the section of the country the coverage has varied from practically no mention to such eye-catching headlines as "Scientist Warns of DDT Peril to Sex Life" while upon further reading you find that the leadoff sentence states "If the pesticide DDT is playing havoc with the sex hormones of rats, rabbits, pigeons, and Japanese quail, what is it doing to man?" or "Sweden Bans DDT Following Reports of Harm to People."

A statement by the National Poisons and Pesticide Board of Sweden on March 27, 1969 presented the following decisions of the Board:

1. Every use of aldrin and dieldrin will be forbidden from the first of January 1970.
2. The use of DDT and lindane in household preparations and similar preparations for home gardening will be forbidden from the same date.

3. Other uses of DDT will be forbidden for a test period of two years from January 1, 1970. In exceptional instances there will be a possibility of an exemption.

The Board assumed that during the test period research will be conducted aiming at getting the effect of the restriction elucidated. The Swedish scientific community has indicated that such studies will be undertaken.

At a conference held in Sweden preceding the announcement it was clearly stated that there is no immediate health risk to man due to the present use of chlorinated hydrocarbon pesticides in Sweden. It was also stated that there exist some risks to the environment. For example, levels of DDT that are of concern are being found in fish from the Baltic Sea far from areas where DDT has been known to be used. Portions of the environment of Northern Sweden have higher DDT residue levels than similar portions of Southern Sweden where DDT has been used.

The members of the conference recognized the importance of having an insecticide which is safe enough to be handled by workers with limited training but which has the required effectiveness for use in campaigns against malaria and other vector-borne diseases in those areas where such diseases are of importance and must be controlled. The World Health Organization was quoted as stating that it is not possible for the time being to do without DDT.

IV. Pesticides and agricultural production

We have mentioned some of the aspects of pesticide pollution and the concern for the environment that Agriculture shares with others interested in our natural resources. We have also pointed out that there are programs suggested and underway to look into the problems associated with pesticide pollution. We previously mentioned that only a small portion of the agricultural lands are treated in any one year but that with certain crops the areas are treated several times in one year. Let us look at agriculture and agribusiness to obtain an estimate of where it stands in the Nation's economy.

Agriculture is the Nation's biggest industry with assets of better than $284 billion. This is equal to about two-thirds the value of all current U. S. corporate assets. Farming involves 5.2 million workers—more than the combined work force needed to operate the Nation's transportation, public utilities, steel and automobile industries. Some 3.2 million people are independent agricultural producers. Three out of every ten jobs in private enterprise are related to agriculture. Eight million people have jobs just storing, processing, and merchandising agricultural products. Six million people work at providing the supplies farmers use.

Agricultural exports also have an impact on the economy. One out of every four acres of land our farmers harvest produces crops for export. The value of U. S. agricultural exports in fiscal year 1968 was about $6.3 billion. Our exports earned more than $5.1 billion worth of dollar exchange and greatly helped the Nation's balance of payments in fiscal year 1968.

The United States is the world's largest exporter of farm products. Our Nation accounts for about 20 percent of the world agricultural trade. Agriculture is big business and any activities such as those illustrated by the headline cited can adversely affect this business either at home or abroad.

Careful study of such charges are needed to determine their validity. Concentrated efforts are also required to reduce the occurrences of pesticide pollution, where it is present. Constant vigilance must be the watchword in the use of pesticides.

V. Controlling pesticides in use

What can be done to control better the role of pesticides in environmental management? One of the main tools at our disposal is an effective information and education program. These programs are well documented and are being used by the Extension Services of our States, by the Federal Extension Service and other government agencies, and by the educational arms of the producers of pesticides and the farm organizations. We feel that the overall program has been effective. We recognize that the program can be more effective. A correctly informed public will provide the necessary balance in the formation of programs of environmental management.

Education and information programs are based on facts. Additional facts regarding the control of insects and the proper usage of pesticides are needed. Continued research in fundamental biology will further elucidate the weak points in the life cycles of the pests and permit new methods of attack in control. We need to continue our studies on safer and more effective use patterns, methods of application, and means of measuring pesticide residues in plants, animals, and their products, and in other parts of the environment. In addition to being able better to measure these residues, we are in need of better information on the significance and effects of these residues in man, his animals, both domestic and wild, the air, water, and soil, and other segments of the environment.

It is generally true at present that we can measure residues of pesticides and their breakdown products at levels which we are unable to understand from the standpoint of effect on the environment.

I believe it is generally agreed that pesticides will be required for the foreseeable future. Extensive progress has been made in the search for means of pest control that is not entirely dependent on chemicals. This research should continue. The development of large-scale area control procedures that utilize the most favorable aspects of the various means of control will aid materially in the reduction of pollution and the management of our environment. Much has been done and much more needs to be done.

I would like to present to you the USDA Policy on Pesticides as outlined in a memorandum from the Secretary of Agriculture. This policy states:

> "It is the policy of the Department of Agriculture to practice and encourage the use of those means of effective pest control which provide the least potential hazard to man, his animals, wildlife, and the other components of the natural environment.
>
> For the foreseeable future, pesticides will be necessary tools for the protection of the nation's food and fiber supplies, people, and their homes.
>
> Where chemicals are required for pest control, patterns of use, methods of application and formulations which will most effectively limit the impact of the chemicals to the target organisms shall be used and recommended. In the use of these chemicals, the Department has a continuing concern for human health and well-being and for the protection of fish and wildlife, soil, air, and water from pesticide contamination.
>
> In keeping with this concern, persistent pesticides will not be used in Department pest control programs when an effective, nonresidual method of control is available. When persistent pesticides are necessary to combat pests, they will be used in minimal effective amounts, applied precisely to the infested area, and at minimal effective frequencies.
>
> "Nonchemical methods of pest control, biological or cultural, will be used and recommended whenever such methods are available for the effective control or elimination of target pests. Integrated control systems utilizing both chemical and nonchemical techniques will be used and recommended in the interest of maximum effectiveness and safety.

In carry out its responsibilities, the Department will continue to:

> Conduct and support cooperative research to find new, effective biological, cultural, and integrated pest control materials and methods;
>
> Seek effective, specific, nonpersistent pesticides and methods of application least hazardous to man and his environment;
>
> Require pesticide product labels which adequately inform all users of the composition and the proper and permitted use of each formulation;
>
> Review and update all pesticide registrations, eliminating any uses not in conformity with current criteria of safety and efficacy;
>
> Cooperate with other public and private organizations and industry in the development and evaluation of pest control materials and methods, assessment of benefits and potential hazards in control operations, monitoring for pesticide residues, and dissemination of pesticide safety information.
>
> All users of pesticides, whether in the home, garden, field, forest, or aquatic area or for public health and sanitary purposes, are strongly urged

to heed label directions and exercise constant care in pesticide application, storage, and disposal for the protection of people, animals, and our total environment.

The Department commends this policy to all who use, recommend, or regulate pesticides."

Summary

Pesticides under controlled conditions are and will be necessary to enable man to better control portions of his environment. Pollution by pesticides in nontarget areas has occurred. In general, levels of pesticides in the food supply are approximately one-tenth of the established allowable daily intake. Levels of pesticides in other portions of the environment are generally low. However, their significance is not well understood.

Alteration of our environment for the benefit of man has produced desirable and undesirable effects. The changes due to pesticides have become strong public issues at local, State, and National levels.

The major tools at our command for pollution management are the information and education programs. A correctly informed public will provide the needed balance in judgment necessary for adequate pollution management.

Résumé*

Le rôle des pesticides dans la maîtrise de la pollution

Dans des conditions contrôlées, les pesticides sont et seront nécessaires pour permettre à l'homme de mieux maîtriser des parties de son milieu ambiant. De la pollution par les pesticides s'est produite dans des régions non visées. En général, la teneur en pesticides de l'alimentation est approximativement de un dixième de la dose quotidienne établie comme admissible. Les quantités de pesticides d'autres parties du milieu ambiant sont généralement faibles. Cependant, leur signification n'est pas bien comprise.

L'altération de notre milieu ambiant a produit au profit de l'homme des effets souhaitables et indésirables. Les modifications dues aux pesticides sont devenues matière à sérieuses discussions à l'échelon local, de l'Etat et du Pays.

Les principaux moyens dont nous disposons pour maîtriser la pollution sont les programmes d'information et d'éducation. Un public correctement informé assurera l'équilibre requis dans l'appréciation nécessaire à maîtriser la pollution de façon appropriée.

---

* Traduit par S. Dormal-van den Bruel.

Zusammenfassung*

Die Rolle der Pestizide in der Kontrolle der Verunreinigungen

Unter kontrollierten Bedingungen sind Pestizide notwendig und werden es auch in Zukunft sein, um dem Menschen eine bessere Kontrolle von Teilen seiner Umwelt zu ermöglichen. Eine Verunreinigung nicht gezielt behandelter Flächen durch Pestizide ist vorgekommen. Im allgemeinen betragen die Pestizidmengen im Lebensmittelangebot etwa $1/10$ der festgelegten zulässigen täglichen Aufnahme. Die Pestizidmengen in anderen Teilen der Umwelt sind im allgmeinen niedrig. Jedoch wird ihre Bedeutung nicht klar erkannt.

Die Änderung unserer Umwelt zum Nutzen des Menschen hat wünschenswerte und unerwünschte Effekte hervorgerufen. Die Veränderungen infolge der Pestizide führten zu strengen öffentlichen Entscheidungen auf lokaler, staatlicher und nationaler Ebene.

Die uns verfügbaren grösseren Hilfsmittel, um mit den Verunreinigungen fertig zu werden, sind Information und Erziehungsprogramme. Eine richtig informierte Öffentlichkeit wird das erforderliche ausgewogene Urteil liefern, das für eine geeignete Behandlung der Verunreinigung nötig ist.

* Übersetzt von F. Bär.

Manuscript received 11 January 1970; accepted 18 February 1970.

Subject Index

GPSR Compliance

The European Union's (EU) General Product Safety Regulation (GPSR) is a set of rules that requires consumer products to be safe and our obligations to ensure this.

If you have any concerns about our products, you can contact us on ProductSafety@springernature.com

In case Publisher is established outside the EU, the EU authorized representative is:

Springer Nature Customer Service Center GmbH
Europaplatz 3
69115 Heidelberg, Germany

Batch number: 10837688

Printed by Printforce, the Netherlands